Thermodynamics Problem Solving in Physical Chemistry

Thermodynamics Problem Solving in Physical Chemistry

Study Guide and Map

Kathleen E. Murphy

CRC Press

Taylor & Francis Group

Boca Raton London New York

CRC Press is an imprint of the
Taylor & Francis Group, an **informa** business

CRC Press
Taylor & Francis Group
6000 Broken Sound Parkway NW, Suite 300
Boca Raton, FL 33487-2742

and by

CRC Press
2 Park Square, Milton Park, Abingdon, Oxon OX14 4RN

International Standard Book Number-13: 978-0-367-23116-3 (Paperback)
International Standard Book Number-13: 978-0-367-23147-7 (Hardback)

Visit the Taylor & Francis Web site at
http://www.taylorandfrancis.com

and the CRC Press Web site at
http://www.crcpress.com

Contents

FINAL ANSWERS

Preface

Many students view physical chemistry problem-solving as largely a mathematical exercise, but it is much more than that. Think of the mathematics as just the means to an end of being able to explain the origin and prediction of the physical properties of chemical substances when by themselves or in chemical mixtures or reactions in all living organisms, pharmacology, engineering and chemistry. Together, the map and workbook can be a real help to you to be successful in applying thermodynamics-based concepts. The purpose and function of the map is to let you know how to "ask the right questions to get the right answers" to a large number of possible situations involving thermodynamic principles, and to extend your analytical thinking and diagnostic abilities. This ability is much more valuable than for just chemistry and will allow you to be more successful in dealing with new situations or problems that you will encounter in your professional career. The map will help you get started but completing the practice exercises in the workbook will help build your confidence and ability to take on different situations. I have deliberately chosen a variety of applications for each major concept so you can see the "ins and outs" of working with the concept or equations. I have also tried to address the need for more visual cues for understanding the problem analysis in the Full Solutions available at https://www.crcpress.com/Thermodynamics-Problem-Solving-in-Physical-Chemistry-Study-Guide-and-Map/Murphy/p/book/9780367231163. The students in my physical chemistry classes have found the map very helpful and I expect you will have the same experience.

Author

I grew up near Saginaw, Michigan and obtained a B.S. degree in Chemistry from Central Michigan University (Mount Pleasant, MI). I then went on to receive a PhD in Physical Chemistry from the University of Vermont (Burlington, VT) and did postdoctoral work at Rensselaer Polytechnic Institute (Troy, NY). I am currently a Professor of Chemistry, in the Natural Science Department at Daemen College (Amherst, NY) and have been teaching general, analytical and physical chemistry for over 40 years. Eighteen of those years were spent being in an administrative role within the department and the College and oversaw the transition of the College into health-related areas and a major expansion of the department. I never left the classroom, though, and look forward to devoting much more of my time using my experience to develop new strategies for students to address their weaknesses in problem-solving and to surmounting them. My research interests have been the role of oxygen in wound healing and various environmental topics, such as bioremediation or detection of metals in the soil or water. My hobbies include gardening, quilting and reading.

Workbook

Gases and Gas Laws

1

KEY POINTS – GAS LAWS

- The equation of state: $PV = nRT$ applies to "ideal gases" where there are either no interactions between particles (at low T's and P's) or where the attractive forces between the gas particles balance the repulsive forces. The equation can be applied to all gases, independent of chemical identity.

Ideal gas:	Van der Waals	Virial equation
$P_{ideal} = \dfrac{nRT}{V}$	$P_{VdW} = \dfrac{nRT}{V - nb} - \dfrac{n^2 a}{V^2}$	$P_{virial} = \dfrac{nRT}{V}\left[1 + \dfrac{nB}{V} + \dfrac{n^2 C}{V^2}\right]$
$P_{ideal} = \dfrac{RT}{V_m}$	$P_{VdW} = \dfrac{RT}{V_m - b} - \dfrac{a}{V_m^2}$	$P_{virial} = \dfrac{RT}{V_m}\left[1 + \dfrac{B}{V_m} + \dfrac{C}{V_m^2}\right]$

- Real gases show deviations from ideal behavior and introduce factors that depend on the chemical identity of the gas molecule.

The **van der Waals equation** introduces two important factors, the values of which depend on the chemical identity of the gas:

- A correction for the molar volume the gas particles themselves occupy, represented by the term "b", which is subtracted from the total molar volume, appears in the first term of the equation.
- The second factor, "a", is a measure of the attractive forces between the gas molecules, represented by the term "an^2/V^2 or a/V_m^2", which is subtracted from the adjusted first term.
- The a and b values are tabled for each gas as van der Waals constants.

The **virial equation** describes gas isotherms as polynomials in V or V_m, where the second virial coefficient B, can be related to "a" and "b" from the van der Waals equation. The value of B is very temperature dependent, different for each gas and may be positive or negative. The third factor, "C", in the equation is rarely needed to define the behavior of a gas.

- The compressibility factor, Z, is a useful parameter with which to determine when a gas is NOT acting ideally, since

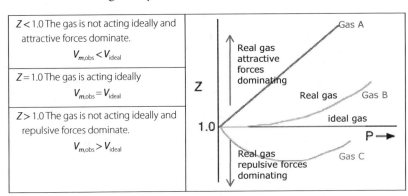

$Z < 1.0$ The gas is not acting ideally and attractive forces dominate. $V_{m,obs} < V_{ideal}$
$Z = 1.0$ The gas is acting ideally $V_{m,obs} = V_{ideal}$
$Z > 1.0$ The gas is not acting ideally and repulsive forces dominate. $V_{m,obs} > V_{ideal}$

$$Z = \frac{P\bar{V}}{RT} = \frac{PV_m}{RT}$$

$$Z = 1 + \frac{B}{V_m}$$

$$Z = 1 + B'P + C'P^2$$

The values of Z vary with T and can be determined by comparing observed values of P, V or other gas properties with those predicted by the ideal gas law.

$$Z = \frac{V_{m,obs}}{V_{m,ideal}} = \frac{\bar{V}_{obs}}{\bar{V}_{ideal}} \quad OR \quad Z = \frac{P_{obs}}{P_{ideal}}$$

The ideal gas law allows for the determination of molecular weights or molar masses of gaseous species, as well as the density of a gas at any P and T combination, if the chemical identity of the gas is known, by defining the number of moles as $n = \dfrac{\text{mass of gas}}{\text{molar mass}} = \dfrac{m}{MW \text{ or } AW}$.

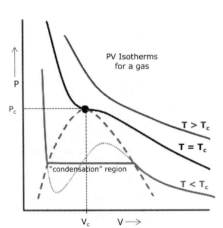

Molar volumes and the density of a gas, both dependent on P and T, are related through the molar mass, which is independent of P and T since:

$$d_{gas}\left(\frac{g}{L}\right) \times V_{m,gas}\left(\frac{L}{mol}\right) = MW\left(\frac{g}{mol}\right)$$

Critical constants for gases: the critical pressure, P_c, and critical temperature, T_c, define the point at which the liquid state ends for a substance. Any gas above the P_c, and T_c cannot be condensed to a liquid, but can be in a "supercritical state". The critical constants for a gas are defined by the van der Waals constants:

$$V_C = 3b \quad P_C = \frac{a}{27b^2} \quad T_C = \frac{8a}{27bR}$$

LAW OF CORRESPONDING STATES

The law states: "all gases with the same reduced pressure and temperature will have the same reduced volume" where the "reduced" variables, are defined as: $P_r = \dfrac{P}{P_C} \quad V_r = \dfrac{V}{V_C} \quad T_r = \dfrac{T}{T_C}$.

- By using reduced variables, the van der Waals equation becomes independent of chemical identity and is expressed as: $P_r = \dfrac{8T_r}{3V_r - 1} - \dfrac{3}{V_r^2}$.

Boyle Temperature: this is the temperature at which a real gas obeys the ideal gas law, over an appreciable P range, which depends on the value of a and b, the van der Waals constants:

$$T_{Boyle} = T_B = \frac{a}{Rb}.$$

EXAMPLE PROBLEMS

1.1   A)   Consider that 25.0 g of $CH_4(g)$ is contained in a sealed 2.0-L container at 30°C. What would be the pressure if the gas were:
(a)   Acting as an ideal gas?
(b)   Acting as a van der Waals gas?
(c)   Obeying the virial equation but only when the first two terms are significant? (at 30°C, B for $CH_4(g) = -43.9$ cm³-mol⁻¹)

   B)   If the volume of the gas was compressed isothermally to 200 mL at $T = 30$°C, what would the pressure in (a), (b), and (c) become?

1.2   Generally it is true that the term involving C (from the virial equation) can be neglected as being too small to affect the final value of the P or Z.

   A)   Given that: $P_{virial} = \dfrac{RT}{V_m}\left[1 + \dfrac{B}{V_m}\right]$ and $Z = \dfrac{PV_m}{RT}$ prove that $Z = \left[1 + \dfrac{B}{V_m}\right]$.

B)   Using the equation for Z involving B, calculate the value of Z for 25.0 g of CH_4 at 30°C in the two different volumes described in Problem 1.1.

C)   Which forces have increased in CH_4 in the smaller volume – the repulsive or attractive forces? Give the reason(s) for your choice.

1.3   A)   Derive the expression for the density of a gas, assuming it is an ideal gas, in terms of:

(a) P, molar mass (MW) and T

(b) MW and V_m

B)   (a)   Define Z in terms of the observed density of a gas compared to the expected ideal gas law value, given that $Z = V_{m,obs}/V_{m,ideal}$.

(b)   What would be true about the observed density of the gas, compared to the expected ideal value, when $Z < 1.0$?

(c)   What would be true about the observed density of the gas, compared to the expected ideal value, when $Z > 1.0$?

1.4   If $Z = 0.8808$ for $CH_4(g)$ at 50°C and 130 atm,

A)   What is the density of the gas, assuming that it obeyed the ideal gas laws?

B)   What is the actual density of the gas?

1.5   In an industrial process, $N_2(g)$ at 300 K and $P = 100$ atm, weighing 92.4 kg, is heated to 500 K in a sealed container with $V = 0.500$ m³. What is its final pressure, assuming that the gas acts:

A)   As an ideal gas?

B)   As a van der Waals gas?

1.6   A)   Given the data for NH_3 gas in the table on the right:

(a)   What is Z for the gas at each temperature and pressure?

(b)   What effect does the increase in pressure have on Z? Which forces are being increased – repulsion or attraction?

	P (atm)	T (K)	$V_{m,obs}$ (cm³/mol)
I	10	340	2606
II	25	340	908.2

B)   Estimate the Boyle Temperature for NH_3.

1.7   Using the van der Waals constants for CO_2, N_2, O_2, and Ar gases:

A)   What value of R should you use for the calculation of the Boyle temperature from the van der Waals constants? Explain your choice.

B)   Calculate the expected Boyle temperature for each of these gases and compare it with the observed temperatures given below.

Gas	CH_4	N_2	H_2	Ar
Observed Boyle temperature	510 K	327 K	110 K	411.5 K

1.8   The density of an ideal gas is found to be 1.8813 g/L at 25°C and 1.00 atm pressure. What is the molar mass of the gas?

1.9   At 500°C and 699 torr, the density of sulfur gas is 3.71 g/L. A molecular form of sulfur exists in the gas phase at this T, as S_n.

A)   What is the value of n?

B)   What is the correct molecular formula for the gaseous species?

1.10   The density of a gas in a closed container is measured as 3.864 g/L at 1.35 bar and 25°C.

A)   If the gas were a diatomic gas, what is the identity of the gas?

B)   If the gas were a mixture of Xe and Ar, what is the % by mass of Ar in the mixture?

1.11 A)   What are the appropriate units for P_r, V_r and T_r?

B)   For the van der Waals equation, written in reduced variables, what are the units on the first and second terms?

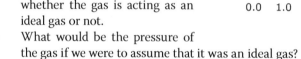

C)   If a gas has a molar volume of 15.0 L/mol at 27°C, what would be the value of its reduced pressure if its critical values were $P_c = 42.7$ atm, $T_c = 151.5$ K and $V_c = 0.0752$ L/mol?

D)   Given the graph of Z versus the reduced pressure on the right, estimate the value of Z and state whether the gas is acting as an ideal gas or not.

E)   What would be the pressure of the gas if we were to assume that it was an ideal gas?

1.12 Given that of a gas has the constant values of $a = 0.751$ atm L²-mol⁻², and $b = 0.0226$ L-mol⁻¹:

A)   Estimate the critical constants P_c, V_c, and T_c for the gas.

B)   What would be the value of Z_c?

1.13 A certain gas obeys the van der Waals equation with $a = 0.500$ Pa m⁶-mol⁻², when its molar volume is 5.00×10^{-4} m³-mol⁻¹ at 273 K and 3.00 MPa.

A)   From this information, calculate the value of b for the gas.

B)   What is the value of Z for the gas under these conditions?

1.14 Could 131.0 g of Xe(g), that was in a sealed container with a volume of 1.0 L at 25°C, exert a pressure of 20 atm if:

A)   It acts as an ideal gas?

B)   It acts as a van der Waals gas?

C)   Considering the same mass of Xe(g) in the sealed container, at what temperature, in °C, would Xe(g) exert a $P = 20$ atm if acting as:

(a)   An ideal gas

(b)   A van der Waals gas?

1.15 An ideal gas undergoes an isothermal compression that reduces its volume by 2.20 L. Its final pressure is 3.78×10^3 torr and its final volume is 4.65 L.

A)   What was the initial pressure of the gas in atm, assuming the gas acts as an ideal gas?

B)   If we had assumed the gas was acting as a van der Waals gas, could we determine the original pressure from the data given? If yes, then determine the pressure. If not, explain why you can't determine the original pressure.

1.16 It is very easy to estimate the molar volume of a gas if it is acting as an ideal gas, with no intermolecular interactions, but not if it is under conditions that produce real gas behavior. For example, the van der Waals equation will become a cubic equation if rearranged to solve for V_m.

$$P = \left[\frac{RT}{V_m - b} \right] - \frac{a}{V_m^2} \quad \Rightarrow \quad V_m^3 - \left(b + \frac{RT}{P} \right) V_m^2 + \frac{a}{P} V_m - \frac{ab}{P} = 0$$

Software is available on the internet that can be used to solve for the correct cube root of the equation. (Resources such as the "cubic equation solver" at CalculatorSoup.com in the algebra section will solve for the three roots of the cubic equation.)

A)   Prove that the van der Waals equation, when rearranged, becomes:

$$V_m^3 - \left(b + \frac{RT}{P} \right) V_m^2 + \frac{a}{P} V_m - \frac{ab}{P} = 0$$

B)   Cylinders of compressed O_2 gas are typically filled to a pressure of 200 bar at 25°C. What would be the molar volume of the $O_2(g)$ if the gas were acting as an:
(a)   An ideal gas?
(b)   A van der Waals gas?

C)   Based on the difference in the values for molar volume in A and B, do you expect that $O_2(g)$ is acting as an ideal gas or a real gas in the tank? Briefly, give the reason(s) for your choice.

1.17   It is not difficult to define the density of a gas using the ideal gas law. However, it is more difficult for real gases, unless Z is known (Problem 1.3). Show that it is possible to find the density by deriving an appropriate equation for the density of a real gas using:

A)   A modified van der Waals equation obtained by assuming $V - nb \approx V$ for the conditions of the gas so that we can start from: $P_{VdW} = \dfrac{nRT}{V} - \dfrac{an^2}{V^2}$

B)   The virial equation, as $P_{Virial} = \dfrac{nRT}{V}\left[1 + \dfrac{nB}{V}\right]$

C)   Compare the values of density you would calculate using the ideal gas law and the equations for real gases derived in (A) and (B) for d_{gas} for $Cl_2(g)$ at $P = 1.00$ atm and $T = 300$ K.

KEY POINTS – MIXTURES OF GASES

▣   Gases in a mixture, that occupy the same volume at the same temperature, will each have "partial pressures" that can be summed to obtain the total pressure of the mixture (Dalton's Law of Partial Pressures): $P_{Total} = P_1 + P_2 + P_3 + \cdots + P_x$

▣   Gases do not need to be acting ideally, but always expect partial pressures to be summed.

▣   Use the term "partial pressure" to reflect that the gas is part of a gaseous mixture.

▣   If each gas obeys the ideal gas law $P_1 = \dfrac{n_1 RT}{V} \rightarrow P_x = \dfrac{n_x RT}{V}$ then the mole fraction, χ, as a pressure ratio can be defined by: $P_{total} = n_{total}\left(\dfrac{RT_{mix}}{V_{mix}}\right)$ and $P_1 = \dfrac{n_1}{n_{total}}(P_{total}) = \chi_1 P_{total}$

▣   If gases are at the same P and T, as in a chemical reaction, $V_{Total} = V_A + V_B + V_C + V_D$ then the volumes add: $V_{Total} = V_1 + V_2 + \cdots + V_x$ $V_{mix} = n_{total}\left(\dfrac{RT_{mix}}{P_{total}}\right)$ and $\dfrac{n_A}{n_{total}} = \chi_A = \dfrac{V_A}{V_{total}}$

▣   For any two gases in the reaction, the stoichiometric ratio is defined by the volume ratio of the reaction volumes needed or the product-to-reactant ratios [Law of Combining Volumes]:

$$\frac{n_A}{n_B} = \frac{a}{b} = \frac{V_A}{V_B} \quad \text{OR} \quad \frac{n_D}{n_B} = \frac{d}{b} = \frac{V_D}{V_B}$$

EXAMPLE PROBLEMS

1.18   If 250 g of Cl_2 and 150 g of C_2H_6 gases are added to a 10-L container at 30°C, what is the total pressure if both gases are acting as:
A)   Ideal gases?
B)   van der Waals gases?

1.19   Carbon monoxide competes with O_2 for binding sites on hemoglobin, but, once bound, CO molecules are not as easily removed as are O_2 molecules, so that CO poisoning is really a lack of O_2 or suffocation. A safe level of CO is 50 ppm, with symptoms of poisoning appearing at 800 ppm, and death occurring within 30 minutes if the level

reaches 3200 ppm. Taking the partial pressure of O_2 as 0.20 in air, what is the ratio of molecules of O_2 to CO when the level is 50 ppm, 800 ppm and 3200 ppm CO?
[Note: 1 ppm = 1 g pollutant per cubic meter of air]

1.20 In normal respiration, an adult exhales about 500 L per hour. If the exhaled air is saturated with water vapor at 37°C, where the equilibrium vapor pressure is 0.062 atm, what mass of water is exhaled in one hour?

1.21 The total pressure of a mixture of O_2 and H_2 gases is 1.00 atm. The mixture is ignited and liquid water produced, which is then removed. The remaining gas is pure H_2 and has a $P = 0.400$ atm when measured at the same V and T of the original mixture. What was the composition of the original mixture in mole percent?

1.22 A rigid vessel with a volume of 5.0 L that contains 500 g of $CO_2(g)$ at 20.5°C is connected through a stopcock valve to a second rigid vessel with a volume of 500 cm³, that contains 35.5 g of Ar(g) at 20.5°C. After the valve is opened to allow the gases to mix:
   A)   If the gases act ideally, what is:
      (a)   The mole fraction of CO_2 in the mixture after opening the valve?
      (b)   The partial pressure of each gas in the mixture?
      (c)   The total pressure in the vessels after the valve has been opened?
   B)   What would have been the answers to (b) and (c) if you consider the gases to be van der Waals gases instead of ideal gases?
   C)   One of the gases does not significantly change pressure when treated as a van der Waals gas. What would be the main factor(s) that make it stay unchanged?

1.23 A gas mixture containing 5 mol % butane, C_4H_{10}, and 95 mol % Ar (such as is used in Geiger-Müller counter tubes) is to be prepared by allowing the gases to fill a 40.0-L evacuated cylinder at 1 atm pressure. The cylinder is then weighed after each gas is introduced.
   A)   Calculate the mass of each gas that needs to be added to produce the desired composition when the temperature is maintained at 25.0°C.
   B)   Determine the density of the final mixture of gases.

1.24 An ideal gas mixture contains 320 mg $CH_4(g)$, 175 mg Ar(g), and 225 mg Ne(g) at 300 K. If the partial pressure of Ne is 66.5 torr in the mixture, what is:
   A)   The total pressure in torr?
   B)   The total volume of the mixture?

1.25 Mount Pinatubo erupted in 1991 with the second largest volcanic eruption in the 20th century. It spewed 20 million tons of $SO_2(g)$ into the high atmosphere where the $T = -17$°C. If the volume of the atmospheric layer it entered (the troposphere) is approximately 8.0×10^{18} m³, what was:
   A)   The partial pressure of SO_2 in the atmosphere?
   B)   The mole % of SO_2 in the atmosphere?

1.26 A mixture of HCl(g), $H_2(g)$, and Ne(g) used for lasers is 5.00% (w/w) HCl, and 1.00% (w/w) H_2, with the rest being Ne. The mixture is sold in cylinders that have a volume of 49.0 L and $P = 138.0$ kPa.
   A)   What is the mole fraction of each gas in the mixture?
   B)   What is the pressure of each gas in the mixture?
   C)   What is the total mass of gas in the flask?

1.27 Suppose the apparatus on the right contains three different gases that have the conditions given on the diagram. Assuming the volume of the connecting tubes to be negligible, compared with the bulb volumes, what would be:

A)   The pressure in each bulb after the valves are opened and equilibrium is established, if the gases act as ideal gases?

B)   The mole fraction of each gas in Bulb A?

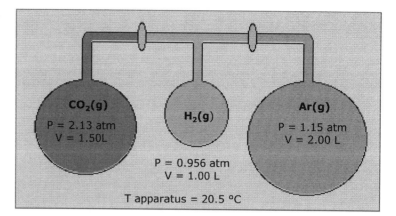

1.28 Given that dry air, which is 78.08% by mass $N_2(g)$, and 20.95% by mass $O_2(g)$ (with the rest being other gases), and has a density of 1.146 g/L at 740 torr and 27°C, determine :

A)   The partial pressure of $N_2(g)$ and its mole fraction in the air.

B)   The partial pressure of $O_2(g)$ and its mole fraction in the air.

1.29 Ammonium nitrate, $NH_4NO_3(s)$, which is used as a fertilizer, is very explosive at high temperatures due to its decomposition reaction: $2\ NH_4NO_3(s) \rightarrow 2\ N_2(g) + 4\ H_2O(g) + O_2(g)$. How many liters of gas would be formed by the decomposition of a 40.0-lb bag of $NH_4NO_3(s)$ at 450°C and 736 torr from 1.0 bar?

1.30 The reaction of sodium peroxide, Na_2O_2, with $CO_2(g)$, is used in space vehicles to convert exhaled $CO_2(g)$ to $O_2(g)$ by the reaction: $2\ Na_2O_2(s) + 2\ CO_2(g) \rightarrow 2\ Na_2CO_3(s) + O_2(g)$.

A)   Assuming that air is exhaled at an average rate of 4.50 L per minute at 25°C, 736 mm Hg, and that the exhaled air is 3.4% $CO_2(g)$ by volume, what is the volume of CO_2 produced daily in the spacecraft?

B)   What volume of O_2 would be produced daily?

C)   How many kilograms of Na_2O_2 would be needed for a year of conversion?

The First Law of Thermodynamics

Work (PV) and Heat, as ΔU and ΔH

2

KEY POINTS – THE FIRST LAW

■ The change in internal energy of a system can be defined as the sum of the changes in work, *dw*, and thermal energy, heat, *dq*.

$$U = q + w \rightarrow dU = dq + dw$$

■ Internal energy, *U*, is a state function and its magnitude does not depend on the pathway, but only on the initial and the final state.

■ Both *dw* and *dq* are pathway dependent, so we must have functions to define and integrate the function from initial to final state, to determine the sign and magnitude of work and heat.

■ Both *dw* and *dq* are negative in sign if they flow out of the system during the change and positive in sign if they flow into system during the change.

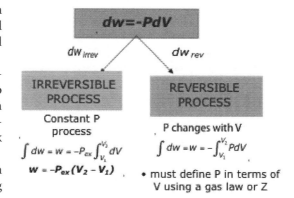

■ In terms of work, pressure–volume (*PV*) work is considered at first, that is: work due to contraction (+), for work done on the system by the surroundings, or due to expansion (−), for work done on the system by the surroundings, and is defined as: $dw = -PdV$.

■ There are two kinds of work, either irreversible or reversible, which produce different functions to evaluate (see map).

■ Thermal energy changes, *q* or heat, have many different definitions, depending on the conditions, but can be divided into the broad categories of isothermal changes involving enthalpy, *dH*, and non-isothermal changes, where kinetic energy and temperature changes occur, which involve a heat capacity, *C*.

■ The relationships and the math that follows, to determine the magnitude of *q*, are summarized on the map, with the main criteria given.

■ For temperature changes, it is important to assess which conditions, constant pressure or constant volume, apply when the system involves gases.

■ The change in internal energy under constant-volume conditions involves no *PV* work, so then $dU = dq_V$. The heat capacity measured under constant *V* conditions, C_V is used to calculate q_V, as shown in your textbook and on the map.

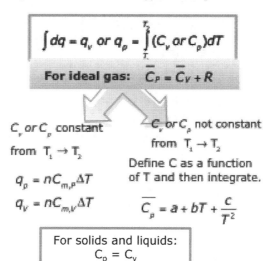

■ Enthalpy, H is defined as: $H = U + PV$ and represents changes in thermal energy that are measured under constant-pressure conditions, as q_P. The heat capacity measured under constant P conditions, C_P, is used to calculate q_P.

■ For gases that undergo the same change in temperature, ΔH and ΔU, will not be equal to each other since $C_P \neq C_V$.

Later in this section, the criteria that apply to ΔH and ΔU for chemical changes under isothermal conditions will be described, but, first, we will deal with situations where no chemical changes are occurring.

EXAMPLE PROBLEMS

2.1   The diagram on the right represents a series of five changes of pressure and volume for 1.0 mol of an ideal gas trapped in a container with a movable piston.

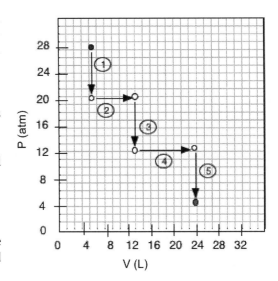

   A)   State in which steps:
   (1)   Work will be done.
   (2)   Temperature of the gas must change.
   (3) ΔU would be zero

   B)   If the gas were He(g), what would be the:
   (1)   Total work done?
   (2)   Total q for the changes?
   (3) ΔU total for the changes?

   C)   If the initial temperature of the gas was 250°C, what will its final temperature be in °C?

2.2   One mole of Ar(g) gas, initially at 20°C and 1000 kPa, undergoes a two-step transformation. For each step of the transformation, calculate the final P, q, w, and ΔU.

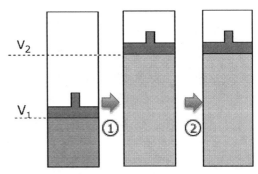

   STEP 1: The gas is expanded isothermally and reversibly until the volume doubles.
   STEP 2: Then, the temperature of the gas is raised to 80°C at a constant volume.

2.3   Two moles of an ideal gas are expanded isothermally and irreversibly from 20 L to 30 L against a constant external pressure. The total work for the process is found to be −5065.8 joules.
   A)   Determine the external pressure in atm present during the expansion.
   B)   Compute ΔU, ΔH, and q for the process.
   C)   If the system is at equilibrium at the end of the expansion, what is the temperature of the system at that point?

2.4   Suppose you started with 2.0 moles of $O_2(g)$ at 273 K with a volume of 11.35 dm³(L)
   A)   What would the ΔU be if the gas temperature was increased to 373 K reversibly and under constant V conditions?
   B)   What would the ΔH be if the gas temperature was increased to 373 K reversibly and under constant P conditions ($P = 1$ atm)?
   C)   Prove that the difference between ΔU and ΔH can be explained if you assume that the gas is ideal, and substitute $\Delta(nRT)$ for $\Delta(PV)$ in the definition: $\Delta H = \Delta U + \Delta(PV)$.

2.5 If 3.0 moles of an ideal gas are compressed isothermally from 60.0 L to 20.0 L, with a constant external pressure of 5.00 atm, what would the values of w, q, ΔU, and ΔH for the compression?

2.6 Consider the isothermal expansion of 5.25 mol of an ideal gas at 450 K from an initial pressure of 15 bar to a final pressure of 3.50 bar, taking into account the figures on the right.
   A) Describe the process in terms of what changes in P and V must be carried out that would result in the greatest amount of work being done by the system and calculate this maximum amount of work done.
   B) Describe the process in terms of what changes in P and V must be carried out that would result in the least amount of work being done by the system and calculate that work done.

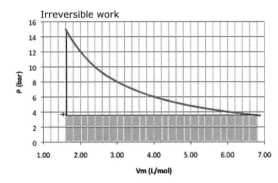

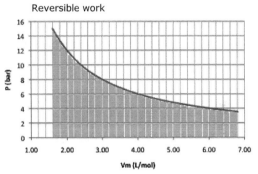

2.7 A sample of blood plasma occupies 0.550 L at 0°C and 1.03 bar. The plasma can be compressed isothermally by a constant pressure of 95.2 bar to a volume that is 94.5% of the original volume.
   A) Treating the plasma as if it were acting as a gas, how much work was done by the surroundings to compress the cells?
   B) If the same amount of work was applied to the same volume of a gas such as $N_2(g)$, starting at the same pressure and temperature, that was compressed with a constant pressure of 10 atm, what would the final volume be?

2.8 Determine the difference in the amount of reversible work that could be done by the expansion of 2.0 moles of $H_2(g)$ at 273 K if it were treated as a real (virial) gas instead of an ideal gas, given $B = 13.7$ cm³/mol for $H_2(g)$ at 273 K.

2.9 Suppose 2.25 moles of an ideal gas at 35.6°C expands isothermally from 26.0 dm³ to 70.0 dm³.
   A) What was the initial pressure of the gas in bars?
   B) Calculate the work done if the expansion occurs against a constant pressure of 0.825 bar.
   C) Calculate the work done if the expansion is accomplished reversibly and isothermally.
   D) Prove that, for an ideal gas, $w_{rev} = -nRT \ln \dfrac{P_1}{P_2}$ and then show that it produces the same magnitude and sign of reversible work as in question C).

2.10 When 229 J of energy is supplied as heat to 3.0 mol of an ideal gas at constant P, the T increases by 2.55 K. Calculate the:
   A) $C_{P,m}$ for the gas.
   B) $C_{V,m}$ for the gas.

2.11 Consider the two situations in the figure on the right. Both Apparatus A and B contain equal masses of ethane, $C_2H_6(g)$, and $O_2(g)$ in sealed containers with pistons. Flask A has a mass M sitting on its piston, while B has two stops on the side of the flask that prevent the piston from moving. Suppose complete combustion takes place at a constant temperature.

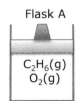

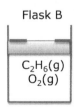

   A) For Flask A:
      (a) Will the piston move from its original position after the reaction is complete? Explain your answer.
      (b) If the piston moves, will the motion be up or down, compared to its original position?
      (c) Can the pressure stay constant during the process, as well as temperature? Explain the reasoning for your choice.

B) For Flask B:
    (a) What will happen to the pressure in the flask before and after the reaction is complete?
    (b) If the heat flow out of Flask B was measured, how would it compare to the heat flow measured for Flask A?
C) How would the ΔU for the two chemical changes in the flasks compare?

ADIABATIC PROCESSES

ADIABATIC PROCESS: $\Delta U = dw$

$\Delta V(+)$ then $\Delta T(-)$

- gas cools when it expands

$\Delta V(-)$ then $\Delta T(+)$

- gas heats when compressed

In **adiabatic** processes, there is no heat flow into the system from the surroundings ($q=0$). Typically, either there isn't enough time for a compensating heat flow in or out of the system for work done on or by the system or the system is thermally insulated from the surroundings. As a consequence, an adiabatic process cannot be isothermal. This condition will impact the amount of work that is either lost or gained by a gaseous system for a given volume change.

The temperature change for a gas undergoing an adiabatic expansion is also associated with the **Joule Thomson Coefficient, μ_{JT}.**

The sign of the Joule Thomson coefficient applies to real gases undergoing an expansion or contraction in an adiabatic process and determines the sign of the T change.

$$\mu_{JT} = \left[\frac{dT}{dP}\right]_H$$

ADIABATIC PROCESS
- must define dT

$dU = dw$
$C_V dT = -P dV$

Reversible process   Irreversible process

- *Must define function for P*
- *If ideal gas, then*

$$C_V dT = \frac{-nRT}{V} dV$$

$$\int_{T_1}^{T_2} \frac{C_V dT}{T} = -nR \int_{V_1}^{V_2} \frac{dV}{V}$$

Assuming C_V constant

$$\overline{C_V} \ln\frac{T_2}{T_1} = -R \ln\frac{V_2}{V_1}$$

$$\frac{T_2}{T_1} = \left[\frac{V_2}{V_1}\right]^{-R/\overline{C_V}}$$

- *Take P as constant*

$$\int_{T_1}^{T_2} C_V dT = -P \int_{V_1}^{V_2} dV$$

- *assuming C_V constant*

$$C_V \Delta T = -P \Delta V$$

Once ΔT defined can calculate work:

$$w_{adiabatic} = nC_V \Delta T$$

For an adiabatic process:

$$\gamma = \frac{C_{P,m}}{C_{V,m}} = \frac{C_P}{C_V} \qquad P_1(V_1)^\gamma = P_2(V_2)^\gamma$$

- For ideal gases, the Joule Thomson coefficient is zero and there is no change from expected behavior given in the box above.
- When a real gas is below its **Joule Thomson inversion temperature**, the coefficient will be positive and the expected change, of cooling when expanding, will occur.
- If, however, the real gas is above its **Joule Thomson inversion temperature,** the coefficient is negative and the gas will warm when it expands, not cool.

This will be explained in more detail in the section before pressure changes and the effect on ΔH is considered.

EXAMPLE PROBLEMS

2.12 A) Calculate the work for the reversible adiabatic expansion of 1.0 mole of an ideal gas with $C_p = 5/2\, R$ at an initial pressure of 2.25 bar from an initial temperature of 475 K to a final temperature of 322 K.

    B) What would the final pressure of the gas have to be if the same work was done but for an isothermal reversible expansion, if T had been kept constant at 322 K and the initial P was 2.25 bar?

2.13 What would the final T be if 2.0 moles $N_2(g)$, initially at 30.0°C, were compressed from 6.0 to 2.0 L:

    A) Adiabatically and irreversibly at a constant P of 1.0 atm?

    B)   Adiabatically and reversibly?
         Furthermore,
    C)   Compare the amount of work required for A) and B).

2.14 If 1.0 mole of an ideal gas, with $C_p = 5/2\ R$, that has an initial pressure of 2.25 bar, goes from an initial temperature of 475 K to 322 K adiabatically:
    A)   Calculate the final volume for the adiabatic expansion
    B)   Calculate the work for the adiabatic expansion in A).
    C)   Calculate the work done if the expansion were a reversible expansion.

2.15 The constant volume heat capacity for a gas and its gamma, γ, the ratio of C_p/C_v, can be measured by observing how T changes when the gas expands reversibly and adiabatically. If the decrease in pressure is also observed, then the C_v can also be determined. Given that a fluorocarbon gas doubles its volume when it is reversibly and adiabatically expanded from 298.15°C to 248.44°C, and the initial pressure was 1522.2 torr:
    A)   Calculate the value of C_v for the gas.
    B)   Calculate the value of gamma, γ, for the gas.

2.16 A)   Given $P_1V_1^\gamma = P_2V_2^\gamma$, prove that $\bar{C}_v \ln\dfrac{T_2}{T_1} = -R\ln\dfrac{V_2}{V_1} \Rightarrow \bar{C}_p \ln\dfrac{T_2}{T_1} = -R\ln\dfrac{P_1}{P_2}$.

    B)   Then, given the final pressure was 693.5 torr for the gas in (2.12), calculate the value of C_p in J/mol K for the gas. How close is this value to the value that could be estimated from γ and C_v calculated in (2.12)?

    C)   If the gas were acting ideally, the final pressure should have been 634 torr. Estimate the value of Z for the gas.

2.17 When air is compressed with a bicycle pump, its temperature rises, so the process is adiabatic.
    A)   If air at 22.0°C and a pressure of 755 torr is pumped into a bicycle tire, with a total volume of 1.0 L, until a pressure of 65 psi is achieved, what would be the final T of the air in the tire, if an adiabatic compression occurs? (14.7 psi = 1.00 atm)
    B)   Using the first law of thermodynamics, explain why the tire heats up.

2.18 A sample of 3.0 mol of an ideal gas at 200 K and 2.0 atm is compressed reversibly and adiabatically until the T increases to 250 K. Given that $C_{V,m} = 27.5$ J/K mol for the gas, calculate:
    A) q
    B) w
    C) ΔU
    D)   the final V of the gas
    E)   the final P of the gas
    F)   and ΔH for the process

KEY POINTS – CALORIMETRY

Calorimeters, whether open (constant P) systems or closed (constant V) systems, are designed to be adiabatic and to not allow heat flow to occur between the system and surroundings. Therefore, the starting premise is that there are at least two parts in the system where the heat loss of one defined part of the system results in an equal gain in another part of the system. So, the mathematical starting point is:

$$-q_{lost} = q_{gain}$$

Based on this situation, you must determine how the loss and gain occurs, whether through chemical change, involving a ΔH or ΔU to define q, or whether

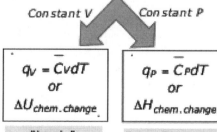
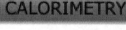

CALORIMETRY

• assume adiabatic
• define two parts, one that loses, the other gains heat
$$-q_{lost} = q_{gain}$$
• assess whether loss or gain occurring through a ΔT or $\Delta U_{chem.\ change}$ or $\Delta H_{chem.\ change}$

Constant V   Constant P

$q_V = \bar{C_V}dT$
or
$\Delta U_{chem.change}$

$q_P = \bar{C_P}dT$
or
$\Delta H_{chem.change}$

"bomb" calorimeter   "styrofoam cup" calorimeter

it undergoes a temperature change, which will involve using C_p or C_v to define q. Substituting correctly for q_{lost} and q_{gain} into the equation will then let you solve for the unknown.

In bomb calorimetry, ΔU_{comb} is measured, not ΔH_{comb}. Then the relationship, based on the gases acting ideally, would be: $\Delta H_{comb} = \Delta U_{comb} + \Delta n(RT)$.

To calculate ΔH you must know Δn for the reaction and assume that the reaction happens at the initial temperature, before the temperature of the water changed, so that you can determine ΔH_{comb}.

EXAMPLE PROBLEMS

2.19 A sample of an organic liquid weighing 0.700 g was burned in a bomb calorimeter, for which the heat capacity of the calorimeter is 2753 J/K and 0.9892 kg of $H_2O(l)$ placed in the calorimeter. The observed temperature change was from 25.00°C to 28.11°C.

   A)   Calculate the ΔU per gram for the combustion.

   B)   If the liquid was propanone, CH_3COCH_3, determine the ΔH_{comb} per mole of propanone from ΔU (assuming $H_2O(l)$ is formed in the combustion). Does it match the tabled value?

   C)   Why did you need to know the identity of the liquid to answer question B), but not A)?

2.20 Seven grams of propene, $C_3H_6(g)$, are burned in a constant-volume calorimeter to form $CO_2(g)$ and $H_2O(g)$ at 25°C. The temperature of the calorimeter and its contents was observed to rise 13.45°C during the process.

   A)   Calculate the heat capacity of the calorimeter and its contents in kJ/K.

   B)   If the state of the water product had been assumed to be liquid, how would the calorimeter constant value change?

2.21 Substances with large positive or negative enthalpies of solution have commercial applications as instant cold or hot packs. Single-use versions of these products involve the dissolution of either anhydrous calcium chloride ($CaCl_2$, $\Delta H_{soln} = -81.3$ kJ/mol) for an instant "hot pack" or ammonium nitrate (NH_4NO_3, $\Delta H_{soln} = +25.7$ kJ/mol) for a "cold pack". Both types consist of a flexible plastic bag that contains liquid water and an inner pouch of the dry chemical. When the pack is twisted and/or struck sharply, the inner plastic bag of solid salt ruptures, and the salt dissolves in the water.

   A)   Considering that 40.25 g of $CaCl_2(s)$ is in the inner package, what would be the temperature change in the 100 mL of water, initially at 22°C, assuming no heat loss to the surroundings, when it dissolves? (density $H_2O(l)$ at 22°C = 0.997 g/mL).

   B)   What is the maximum mass of NH_4NO_3 which should be enclosed in a cold pack containing 100 mL of $H_2O(l)$ at 22°C, if you don't want it to freeze when the solid fully dissolves? Given the solubility of NH_4NO_3 is 119 g per 100 mL of water, would all the calculated amount dissolve?

   C)   Why can this situation be considered to be a constant P process, so ΔH can be used for q, and not a constant-volume process, if the package is sealed?

2.22 Complete combustion of 1.0 L of a natural gas mixture of $CH_4(g)$ and ethane, $C_2H_6(g)$, at 0.0°C and 1.0 atm gives off 43.6 kJ of heat, producing $CO_2(g)$ and $H_2O(l)$. If the product and reactant gases are at same P, T for the combustion, what is the % composition of the mixture by volume?

2.23 Propane gas, $C_3H_8(g)$, is kept in a 200-L tank at a pressure of 2.35 atm at 26°C. Some gas is withdrawn and burned in excess $O_2(g)$ until the tank pressure in the gas drops to 1.10 atm. Some of the heat from the combustion was used to heat 132.5 L of water from 26.0°C to 62.2°C.

   A)   Assuming the combustion products are $CO_2(g)$ and $H_2O(l)$, what is ΔH_{comb} of the gas at 25°C?

   B)   Assuming ΔH_{comb} is constant, what % of the heat from the combustion is used to heat the water?

2.24 Suppose someone had proposed an "ice diet", where energy absorbed by ice melting would consume the excess Calories (1 Cal = 4.184 kJ) from eating too much food. A typical daily intake of energy needs to be about 2500 Calories.

A)   How many kilograms of ice at 0°C would you have to consume and melt at 0°C to consume 500 Calories, if you consider only the melting process?

B)   How would the amount of ice change if you include the heating of the melted water to 37°C? Is it wise to ignore it?

2.25   Lime, CaO, reacts with water to form $Ca(OH)_2(s)$ as the only product:

$$CaO(s) + H_2O(l) \rightarrow Ca(OH)_2(s)$$

When 56.0 g CaO is added to 100 mL $H_2O(l)$ in an insulated container at 26°C and 1 atm, and a complete reaction occurs, if the heat is absorbed by any liquid water left in the container, what will be the number of moles of each substance in the container at the end of the reaction, the final temperature and the state of each substance? (Ignore the solubility of $Ca(OH)_2$ in water and assume that there is no heat lost or gained from the surroundings.)

KEY POINTS – HESS'S LAW CALCULATION FOR ΔH

Enthalpy measures the energy changes when interactions or bonding of atoms change in reactions or other processes. The map lists a number of different types of defined enthalpies and many of these are difficult to measure directly.

One method to determine the enthalpy of a reaction is to use Hess's Law, comparing the final and initial states, using standard enthalpies of formation of substances. It is also possible to use Hess's Law to determine the standard enthalpy of formation, from a measured value for a known reaction involving the substance, by rearranging Hess's Law, as shown on the map (and on the right).

> *HESS'S LAW*
>
> For the general reaction: $aA + bB \rightarrow cC + dD$
>
> $$\Delta H_r^\circ = \left[c\Delta H_{f,C}^\circ + d\Delta H_{f,D}^\circ \right] - \left[a\Delta H_{f,A}^\circ + b\Delta H_{f,B}^\circ \right]$$
>
> can be rarrange to solve for $\Delta H_{f,C}^\circ$:
>
> $$\frac{\Delta H_r^\circ + \left[a\Delta H_{f,A}^\circ + b\Delta H_{f,B}^\circ \right] - d\Delta H_{f,D}^\circ}{c} = \Delta H_{f,C}^\circ$$

EXAMPLE PROBLEMS

2.26   The hydrocarbon liquid, hexadecane, $C_{16}H_{34}$, has a heat of combustion equal to −10,700 kJ per mole at 298.15 K and 1.0 bar. Given the products of the combustion reaction are $H_2O(l)$ and $CO_2(g)$, what would be the:

A)   Standard enthalpy of formation for hexadecane, $C_{16}H_{34}$, in kJ per mole?

B)   The ΔU_{comb} for hexadecane, $C_{16}H_{34}$, at 298.15 K and 1.0 bar?

C)   As an alkane, hexadecane would be close in structure to the lipids or fat biomolecules. Compare the caloric content of hexadecane, $C_{16}H_{34}$, to that of an average fat, which is 9.0 kcal/g.

2.27   The temperature rise of 955 g $H_2O(l)$, measured in a constant-pressure calorimeter, was 7.03°C at 1.0 bar when 10.52 g of solid glucose, $C_6H_{12}O_6(s)$, was completely reacted with oxygen gas, at 22°C and 1.0 bar, so that it produced pyruvic acid, CH_3COCO_2H, and $H_2O(l)$.

A)   Write the balanced reaction occurring.

B)   What is the standard enthalpy of pyruvic acid in kJ/mol from these data? [Given C_p, $H_2O(l) = 4.184$ kJ/°C-kg]

KEY POINTS – INDIRECT METHODS FOR ΔH DETERMINATION

One "indirect" method to find ΔH values that cannot be directly measured is to add several reactions with known values of ΔH together in such a way as to produce the ΔH_r value sought, which is essentially what Hess's Law does. The boxes on the right (from the map) outline the process used.

> CALCULATING AN UNKNOWN ΔH FROM KNOWN VALUES
>
> - Can calculate as a series of stepwise changes
> - ΔH, ΔU are state functions, independent of pathway

STEPWISE ADDITION of KNOWN REACTIONS

$a \times$ (reaction 1)	$a \times \Delta H_1$
$+ b \times$ (reaction 2)	$b \times \Delta H_2$
$+ c \times$ (reaction 3)	$c \times \Delta H_3$

Summed reaction: reactants → products

$\Delta H_{\text{summed reaction}} = a\Delta H_1 + b\Delta H_2 + c\Delta H_3$

To correctly add the reactions, you must first know the balanced reaction you want to determine ΔH_r for. Then, focus on the substances that are "unique" – that is, which appear in the reaction you want and in ONLY ONE of the listed reactions. To get that substance on the correct side of the reaction arrow, you may have to reverse the reaction (which will change the sign on the ΔH for that reaction) and possibly multiply by either a whole number or a fraction in order to match the coefficient for that substance in the overall final reaction you seek.

EXAMPLE PROBLEMS

2.28 The enthalpy of hydration for a salt like sodium acetate trihydrate can be determined by comparing the enthalpy of solution of anhydrous sodium acetate and that for sodium acetate trihydrate solids.

$$NaC_2H_3O_2(s) \xrightarrow{H_2O} Na^+(aq) + C_2H_3O_2^-(aq) \quad \Delta H_{\text{soln}} = -17.32 \text{ kJ}$$

$$NaC_2H_3O_2 \cdot 3H_2O(s) \xrightarrow{H_2O} Na^+(aq) + C_2H_3O_2^-(aq) + 3H_2O(l) \quad \Delta H_{\text{soln}} = 19.66 \text{ kJ}$$

A) Show how to add the reactions to determine the enthalpy of hydration per mole of sodium acetate for the trihydrate form.
B) Then, calculate the enthalpy of hydration in kJ per mole of water for the hydrate.

2.29 Calculate the standard enthalpy for the fermentation of glucose solid, $C_6H_{12}O_6(s)$, which produces $C_2H_5OH(l)$ and $CO_2(g)$: $C_6H_{12}O_6(s) \rightarrow 2C_2H_5OH(l) + 2CO_2(g)$ from the heats of combustion of ethanol liquid and glucose solid.

2.30 The molecule, OH(g), is a free radical that is called the "detergent" molecule of the atmosphere. It reacts with many hydrocarbons or other molecules that enter the atmosphere to oxidize them into forms that can be removed from the atmosphere. It is present in very low concentrations at any time in the atmosphere, so that the thermodynamics of its production are difficult to measure directly.
A) Given the enthalpy of the following reactions, determine the ΔH for the reaction: $H_2O(g) \rightarrow H(g) + OH(g)$ by adding the appropriate reactions.

Reaction 1:	½ H$_2$(g) + ½ O$_2$(g) → OH(g)	$\Delta H = 38.95$ kJ
Reaction 2:	H$_2$(g) + ½ O$_2$(g) → H$_2$O(g)	$\Delta H = -241.8$ kJ
Reaction 3:	H$_2$(g) → 2 H(g)	$\Delta H = 496.0$ kJ
Reaction 4:	O$_2$(g) → 2 O(g)	$\Delta H = 498.3$ kJ

B) In which of the listed reactions would the ΔH be an enthalpy of formation, ΔH°_f, and in which would the ΔH be defined as a $\Delta H_{\text{dissociation}}$ or bond energy?

2.31 The combustion of elemental boron, B(s), produces $B_2O_3(s)$ as the only product and releases 1274 kJ of heat per mole. The combustion of diborane, $B_2H_6(s)$, yields $H_2O(l)$ and $B_2O_3(s)$ and has an enthalpy of –2034 kJ per mole.
A) From these two combustion equations and any other defined reactions with known enthalpies, calculate the ΔH°_f of diborane, $B_2H_6(g)$.
B) Show that solving the problem by using Hess's Law for the combustion and rearranging to solve for the ΔH°_f of diborane, $B_2H_6(g)$, produces the same result.

2.32 Choose the appropriate reactions and then describe the route to calculate the ΔH by adding those reactions to produce the ΔH_r for: $CS_2(l) + 3 Cl_2(g) \rightarrow CCl_4(l) + S_2Cl_2(l)$

Reaction I:	CS$_2$(l) + 3 O$_2$(g) → CO$_2$(g) + SO$_2$(g)	$\Delta H_I = -1077$ kJ
Reaction II:	2 S(s) + Cl$_2$(g) → S$_2$Cl$_2$(l)	$\Delta H_{II} = -58.2$ kJ
Reaction III:	C(s) + 2 Cl$_2$(g) → CCl$_4$(l)	$\Delta H_{III} = -135.4$ kJ

Reaction IV:	$S(s) + O_2(g) \rightarrow SO_2(g)$
Reaction V:	$SO_2(g) + Cl_2(g) \rightarrow SO_2Cl_2(g)$
Reaction VI:	$C(s) + O_2(g) \rightarrow CO_2(g)$
Reaction VII:	$CCl_4(l) + O_2(g) \rightarrow COCl_2(g) + Cl_2O(g)$

$\Delta H_{IV} = -296.8$ kJ

$\Delta H_V = -97.3$ kJ

$\Delta H_{VI} = -393.5$ kJ

$\Delta H_{VII} = -5.2$ kJ

KEY POINTS – USING CYCLIC DIAGRAMS FOR ΔH

A second type of indirect method uses cyclic diagrams to compare the total heat generated in two independent pathways to achieve the same final state from the same initial state to determine an unknown value. This proves extremely useful for solving difficult-to-measure values such as bond energies $(\Delta H_{dissociation})$, as well as a number of other thermodynamic parameters. Cyclic diagrams are a useful approach for determining the change in ΔH_f or $\Delta H_{reaction}$ at T other than 298 K. Again, the map outlines the basic premise that the total heat needed for pathway (1) will equal that of pathway (2), since thermodynamic quantities such as ΔH and ΔU are state functions.

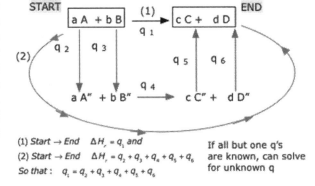

(1) *Start → End* $\Delta H_r = q_1$ and

(2) *Start → End* $\Delta H_r = q_2 + q_3 + q_4 + q_5 + q_6$

So that : $q_1 = q_2 + q_3 + q_4 + q_5 + q_6$

If all but one q's are known, can solve for unknown q

EXAMPLE PROBLEMS

2.33 A)   Using standard heats of formation, calculate the $\Delta H°_{298}$ for the decomposition of hydrogen peroxide, H_2O_2, into water and oxygen gas at 25°C and 1.0 atm:

$$2H_2O_2(g) \rightarrow 2H_2O(g) + O_2(g)$$

	$\Delta H_f°$(kJ/mol)
$H_2O_2(g)^1$	−136.3
$H_2O_2(aq)^2$	−191.7
^{1} CRC Handbook Table	
^{2} ATcT Argonne National Lab	

B)   Construct a pathway that will allow you to estimate the bond dissociation energy for the O-O bond in gaseous hydrogen peroxide.

C)   If an enzyme (catalase) normally converts $H_2O_2(aq)$ with the reaction:

$$2H_2O_2(aq) \rightarrow 2H_2O(l) + O_2(g)$$

does this change the $\Delta H°_{298}$ for the decomposition? Prove your answer.

D)   Why would the enzyme (catalase) reaction not be a useful reaction for estimating the O-O bond dissociation energy? Explain.

2.34 Xenon gas forms $XeF_6(s)$ by reaction with $F_2(g)$. The $\Delta H°_f$ for $XeF_6(s)$ is −363.2 kJ/mol. The compound sublimes with decomposition at 25°C with a $\Delta H°_{sublimation} = 62.34$ kJ/mol. Using this information and the fact that $\Delta H_{dissociation}\ F_2(g) = 158.0$ kJ/mol, construct a pathway that would allow you to determine the bond energy of the Xe-F bond and then estimate its value from the pathway.

2.35 If you were to construct a pathway that could define the standard enthalpy of formation of 1 mole of $NH_3(g)$ in terms of other known enthalpies, as shown on the right:

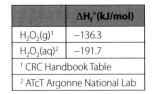

A)   What values would you need to look up for q_1, q_2, and q_3?

B)   Calculate the $\Delta H_f°$ from your pathway and compare to the tabled value of −46.11 kJ/mol.

C)   Bond energies describe the energy needed to break a bond in a gaseous molecule into gaseous single atoms and cannot account for the extra energy that appears as interactions between the atoms, either in the reactant or product state. They are quoted as average energies over many different types of molecules and therefore cannot account for specific interactions (such as interatomic or intermolecular forces, resonance energy, etc.). What major type of interaction could occur between the NH_3 molecules that could explain the discrepancy observed in B)?

2.36 Because bond energies apply to molecules formed from single atoms in the gaseous state, not all pathways can be based on just bond energies. For the pathway describing the formation of octane liquid, $C_8H_{18}(l)$,

$$8\ C\ (s,\ graphite) + 9\ H_2\ (g) \xrightarrow{\Delta H_r^{\circ}} C_8H_{18}(l)$$

$$\downarrow q_1 = ? \qquad \downarrow q_2 = ? \qquad \nwarrow q_4 = ?$$

$$8\ C\ (g) \ + \ 18\ H(g) \xrightarrow[q_3 = ?]{} C_8H_{18}(g)$$

from its elements in their most common state at 298.15 K and 1.0 atm, as shown on the right:

A) What values would you need to look up for $q_1, q_2, q_3,$ and q_4? Use those values to estimate the ΔH_f° for liquid octane.

B) Calculate the ΔH_f° for $C_8H_{18}(l)$, from the heat of combustion which is −5470.2 kJ/mol, when $CO_2(g)$ and $H_2O(l)$ are formed, using Hess's Law. Compare it to the value determined in A).

C) If you assume each C atom in graphite is sp^2, with two single bonds and one pi bond, and used the bond energies for breaking these bonds for q_i, what would the calculated ΔH_f° have been? What would the difference between this calculated value and the value calculated in B) have indicated to you?

D) If we calculate a bond energy value from the pathway approach, why is it always referred to as an "*estimate* of the bond energy"?

2.37 Ionic compounds have an additional stability due to the lattice energy between the ions in the solid structure, which adds to the energy of bond formation. The lattice energy is defined as the energy required to separate an ionic solid into its component monatomic gaseous ions, which cannot be directly measured. But lattice energies can be estimated using cyclic diagrams (called Born–Haber cycles), which use bond energies and other known ΔH values to produce the chemical change needed from the formation equation for the compound.

A) Construct a cycle, using only known ΔH values, that will let you determine the lattice energy of KF(s) from its heat of formation.

B) Use the cycle to estimate the lattice energy for KF(s) and compare it to the tabled values that are between 801 and 822 kJ per mole.

KEY POINTS – TEMPERATURE DEPENDENCE OF ΔH

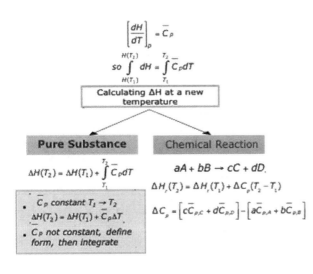

The T dependence of ΔH is defined as the heat capacity measured under constant pressure. Since heat capacities change markedly with T, the values of ΔH are T dependent. But often the ΔH values do not vary greatly, unless the temperature change is very large. The only way to determine how much the ΔH will be affected by a change in T is to calculate the "extra" heat loss or gain from the heat capacity is to integrate (as shown on the map) the known function describing how C_p changes with T: $\int_{T_1}^{T_2} \bar{C_p} dT$

■ The T dependence of $\bar{C}_p$ (or $C_{p,m}$) is often expressed as a polynomial in T or as a power series in T, depending on which equation best fits the C_p versus T data measured.

$$\bar{C}_p = a + bT + \frac{c}{T^2} \ \text{or} \ \bar{C}_p = a + bT + cT^2 \ \text{or} \ \bar{C}_p = A(1) + A(2)T + A(3)T^2 + A(4)T^3$$

(The heading of the table giving the values of the coefficients will show the form of the polynomial they apply to.)

■ When C_p is a power series, the first three terms contribute significantly to the value and none can be ignored.

■ When in the form $\bar{C}_p = a + bT + \frac{c}{T^2}$, the first two terms are the most important in setting the value and the third term may be ignored, unless c is very large.

- The polynomial coefficients may be determined for T in °C, instead of K, so be sure to check the table heading.
- For a chemical reaction, the difference between the heat capacity of products versus reactants is of primary importance and may cause the ΔH for the reaction to change significantly. The change may be estimated from the difference between the $\overline{C}_{p,298}$ values of products and reactants (see map) or, more precisely, from the individual change for each substance's $\Delta H_f°$ before Hess's Law is applied.
- Cyclic diagrams can be very useful for organizing the information needed to calculate the change in q_p for a chemical reaction to achieve an overall change (as shown on the right).
- The effect of changing T on ΔH for a chemical reaction can also be summarized by **Kirchhoff's Law**:

$$\Delta H_r°(T_2) = \Delta H_r°(T_1) + \Delta \overline{C}_p \int_{T_1}^{T_2} dT$$

where ΔC_p for a reaction:

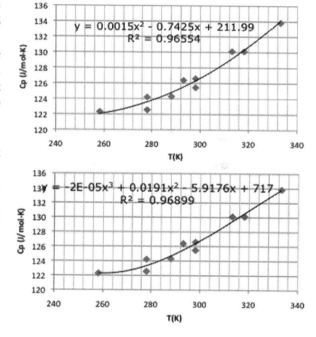

At T2   a A   +   b B $\xrightarrow{q_1}$ c C   +   d D   **ΔH = q₁= ?**

| Cool A, q₂ T2 → T1 | Cool B, q₃ T2 → T1 | Heat C, q₅ T1 → T2 | Heat D, q₆ T1 → T2 |

At T1   a A   +   b B →   c C   +   d D   **ΔH$_{T1}$ = q₄**

ΔH (T2) = q₂+ q₃+ΔH(T1) + q₅+ q₆

$aA + bB \rightarrow cC + dD$ is defined as: $\Delta \overline{C}_p = \left[\left(c\overline{C}_{p,C} + d\overline{C}_{p,D}\right) - \left(a\overline{C}_{p,A} + b\overline{C}_{p,B}\right)\right]$

EXAMPLE PROBLEMS

2.38 The Dortmund Data Bank lists data for the temperature variation of the $C_{p,m}$ for acetone liquid, $CH_3COCH_3(l)$, from 258.15 to 333.15 K.

After plotting the data, the best-fit polynomial could be either to the power of 2 or 3 as shown in the graphs on the right.

A)   If the polynomial $\overline{C}_p = a + bT + cT^2$ is applied (first graph), what would be the units of a, b, and c in the equation?

B)   Compare the results of using the first graph's polynomial $\overline{C}_p = a + bT + cT^2$ to calculate the value of $C_{p,m}$ at 298 K and to the tabled value of 124.3 J/mol-K.

C)   Do the same for the second graph's best-fit polynomial: $\overline{C}_p = A(1) + A(2)T + A(3)T^2 + A(4)T^3$.

D)   Could any of the terms in either polynomial be ignored as being insignificant?

First graph: $y = 0.0015x^2 - 0.7425x + 211.99$ $R^2 = 0.96554$

Second graph: $y = -2E{-}05x^3 + 0.0191x^2 - 5.9176x + 717$ $R^2 = 0.96899$

2.39 Given the values in 2.28, what would be the value of $\Delta H_f°$ for $CH_3COCH_3(l)$ at 400 K:

A)   If you assumed $C_{p,m}$ was constant from $298 \rightarrow 400$ K, using the tabled value, 124.3 J/mol K?

B)   If the polynomial $\overline{C}_p = a + bT + cT^2$ from the first graph in 2.28 replaced C_p in the integral?

2.40 Consider the reaction: $2CH_3OH(g) \rightarrow 2CH_4(g) + O_2(g)$

A)   Calculate $\Delta H°_{298}$ for the reaction using Hess's Law.

B)   To calculate the ΔH_r for a reaction at some temperature other than 25°C, you can develop a cyclic process (as illustrated above), using the heat capacities of each substance (but NOT involving bond energies!) from a known $\Delta H°_{298}$.

(a)   Describe the pathway, identifying the heat terms needed, that would allow you to calculate the $\Delta H°$ of the reaction at 500°C.

(b)   Calculate the value for ΔH_r at 500°C, assuming heat capacities are constant, using your pathway.

T (K)	CH₄	Cp (J/mol-K), CH₃OH (g)	O₂ (g)
298	35.695	44.101	29.378
300	35.765	44.219	29.387
400	40.631	51.713	30.109
500	46.627	59.8	31.094
600	52.742	67.294	32.095

C)   The heat capacity values for reactants and products are shown in the box on the right. When C_p is plotted versus T (K), each substance shows a linear relationship ($y = mx + b$) and the table on the right gives the values of the coefficients.

D)   How much of a difference in the calculated value for ΔH_r at 500°C (773 K) would be observed if the C_p functions are integrated to calculate the q values that were calculated in 2.40 B (b), using the values in the box on the right for a and b and the integrated result of $C_p = a + bT$?

$C_p = a + bT$	a (J/mol-K)	b (J/mol-K²)
$CH_3OH(g)$	22.811	0.0726
$CH_4(g)$	18.347	0.0572
$O_2(g)$	26.677	0.0089

2.41 The heat of formation of tungsten carbide, a network covalent solid, is difficult to measure directly since it only forms at 1400°C with the reaction: $W(s) + C(s)$, graphite → $WC(s)$

A)   Estimate the ΔH for the formation of 1 mol $WC(s)$ at 25°C from the listed reactions below:

$$2W\,(s) + 3O_2\,(g) \rightarrow 2WO_3\,(s) \qquad \Delta H = -1680.6\,kJ$$
$$C\,(graphite) + O_2\,(g) \rightarrow CO_2\,(g) \qquad \Delta H = -393.5\,kJ$$
$$2WC\,(s) + 5O_2\,(g) \rightarrow 2WO_3\,(s) + 2CO_2\,(g) \qquad \Delta H = -2391.6\,kJ$$

B)   Calculate the $\Delta H_f°$ for the formation of $WC(s)$ at 1400°C, assuming that all the C_p values are constant and knowing that the states for reactants and products don't change.

C)   The heat capacity for $C(s)$, graphite changes markedly over the temperature range 500–1400°C. Using the equation for the heat capacity of $C(s)$ graphite given below, recalculate the term for $C(s)$ in (B) and see how much of an effect this would have on the value of the $\Delta H_f°$.

Heat Capacity C(s), graphite			
	a	b	c
$\overline{C}_p = a + bT + cT^2$	−0.4493	0.03553	-1.3×10^{-5}
Source: Handbook Chemistry & Physics, 88th edition			

2.42 Consider a reaction in which the amino acid, aspartic acid, $C_4H_7NO_4$, is converted to alanine, $C_3H_7NO_2$, and $CO_2(g)$.

A)   To determine the ΔH for the reaction at 50°C and 1.0 bar:

(1)   What assumptions would you need to make to be able to estimate ΔH for the reaction at 50°C?

(2)   State what properties of the molecules you would have to know, and

(3)   What equations you would have to use.

Aspartic acid   Alanine

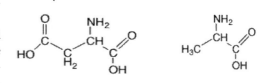

B)   Estimate the ΔH for the reaction at 50°C, using published values from standard sources such as the CRC Handbook of Chemistry and Physics or the National Institute of Standards (NIST) Chemistry web book (https://webbook.nist.gov/chemistry/).

2.43 Under anaerobic conditions, glucose, $C_6H_{12}O_6$, is converted to lactic acid, $C_3H_6O_3$, in biological cells.

$$C_6H_{12}O_6\,(s) \rightarrow 2CH_3CH(OH)CO_2H\,(s)$$

A)   What is the ΔH for the reaction at 25°C and 1.0 bar?

B)   What would be the ΔH for the reaction at 37°C (and 1.0 bar)?

C)   Is this reaction ΔH-sensitive to the small change in temperature incurred at body temperatures?

2.44 The Haber–Bosch reaction: $N_2(g) + 3\ H_2(g) \rightarrow 2\ NH_3(g)$ is a very important way to produce NH_3 for fertilizers. The reaction must be run at temperatures in the range of 400–450°C and at high pressures (200 atm) to increase the yield of the reaction. Dealing with only the temperature change:

A) What would be the ΔH for the reaction at 25°C?

B) Estimate the effect of increasing the temperature to 425°C on the reaction ΔH, assuming all C_p values remain constant from $T_1 \rightarrow T_2$.

2.45 On a table of change-in-state values, the ΔH_{vap} at the boiling point of CH_3OH, 337.2 K is given as 35.27 kJ/mol, but, under the value, you are told that the ΔH_{vap} would be 37.99 kJ/mol at 298 K.

A) Given what you now know about how to calculate the change in a ΔH value at a new temperature, show how you could estimate the ΔH_{vap} at 298 K from the ΔH_{vap} at the normal boiling point and the heat capacities of the liquid and gas states of CH_3OH.

B) Another way to approach the calculation is to use Hess's Law to find ΔH_r for the change-in-state by seeing it as the reaction: $CH_3OH(l) \rightarrow CH_3OH(g)$, providing the respective ΔH_f° values are known. Compare the value you would get from your method in A) to that using Hess's Law, as described.

2.46 One mole of an alkane, $C_nH_{(2n+2)}$, with $C_p = 120.2$ J/mol K is burned in excess $O_2(g)$ to form $CO_2(g)$ and $H_2O(g)$) at 298 K. Under constant conditions, pressure q is measured as −3536.1 kJ. If the same reaction is run at 400°C, the q is measured at −3532.8 kJ. What is the molecular formula of the alkane?

Second and Third Law of Thermodynamics, ΔS

3

KEY POINTS – THE SECOND LAW

Entropy is defined by the Second Law of Thermodynamics and is representative of the arrangement of particles ("order") in the system or the distribution of thermal energy in "microstates" in the system. It is a critical part of chemical thermodynamics and defines the natural direction of changes in the Universe.

■ Entropy is a state function, like ΔU and ΔH, so that ΔS is independent of the pathway and depends only on the final and initial states of the system.

■ The entropy of a substance depends on the temperature and the physical state of a substance. The two basic ways to calculate ΔS for an isothermal process, such as a phase transition, and for a range of temperatures, are given on the map.

■ Unlike enthalpy, where we assume the elemental forms are the zero state, using the Third Law of Thermodynamics, we can define a "zero" state for entropy, $S = 0$, for perfect crystalline materials.

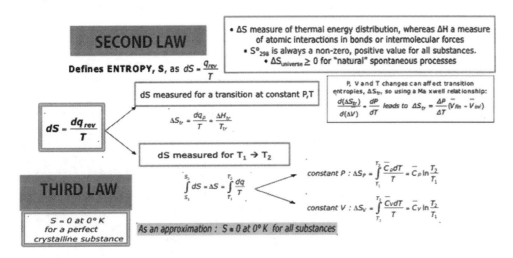

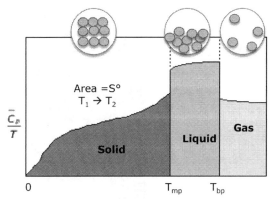

Area = S°
$T_1 \rightarrow T_2$

$\dfrac{\overline{C}_p}{T}$

Solid
Liquid
Gas

0   T_{mp} T_{bp}

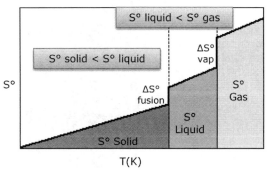

S° liquid < S° gas

S° solid < S° liquid

$\Delta S°$ vap

$\Delta S°$ fusion

S° Gas

S° Liquid

S° Solid

S°

T(K)

- The absolute entropy, $S°$, can be determined from the heat capacity functions for the various phases below 298.15 K, and the ΔH values for all phase changes. The entropy from 0.0 K to about 20 K is estimated using the Debye function. In general, the calculation is described by:

$$\overline{S}°_{T(g)} = \int_0^{T_{mp}} \frac{\overline{C}_{p,X(s)}}{T}dT + \frac{\Delta \overline{H}°_{\text{fus},X(s)}}{T_{mp}} + \int_{T_{mp}}^{T_{bp}} \frac{\overline{C}_{p,X(l)}}{T}dT + \frac{\Delta \overline{H}°_{\text{vap},X(l)}}{T_{bp}} + \int_{T_{bp}}^{T} \frac{\overline{C}_{p,X(g)}}{T}dT$$

If the C_p is defined by the function for a state, $\overline{C}_p = a + bT + cT^2 + dT^3$ then the integrals for C_p/T become:

$$\Delta S = \int_{T_1}^{T_2} \frac{\overline{C}_p}{T}dT = \int_{T_1}^{T_2} \frac{a + bT + cT^2 + dT^3}{T}dT$$

$$= a\int_{T_1}^{T_2} \frac{dT}{T} + b\int_{T_1}^{T_2}dT + c\int_{T_1}^{T_2}TdT + d\int_{T_1}^{T_2}T^2 dT$$

- Every substance, including elemental forms, has a $S°$ value at 298 K and it must always be positive.
- For ΔS in a phase change (for $a \rightarrow \beta$ $\Delta S°_{a \rightarrow \beta} = S°_\beta - S°_a$) or a change in state, the "sign" will be positive or negative, depending on the direction, (–) for cooling, (+) for heating, and will occur at a specific temperature.
- The ΔS associated with a chemical reaction, as an isothermal change, can be either positive, negative or possibly zero and is defined in a manner similar to Hess's Law, as shown on the map.
- The ΔS for a chemical reaction at temperatures other than 298 K can be defined by using a cyclic process or an equation derived from Kirchhoff's Law.

$$\Delta S°_r(T_2) = \Delta S°_r(T_1) + \Delta \overline{C}_p \ln\left(\frac{T_2}{T_1}\right) \text{ where } \Delta \overline{C}_p = \sum_{\text{products}} \overline{C}_p - \sum_{\text{reactants}} \overline{C}_p$$

- Changes in volume will result in non-zero values for ΔS since changes in arrangement will affect the distribution of allowed states for the atoms or molecules. These changes in entropy must be calculated along a reversible pathway. (The appropriate equations are described on the map.)
- A pressure change will also affect the entropy of a gas, which is described on the map, but will have little effect on the entropy of solids or liquids.
- When two properties are changed together, such as: $V_1T_1 \rightarrow V_2T_2$ (or $P_1T_1 \rightarrow P_2T_2$) for a gas, the process must be broken down into two steps, e.g. $V_1T_1 \rightarrow V_2T_1$ followed by $V_2T_1 \rightarrow V_2T_2$, where the entropy change from each step can be added to produce the total entropy change to define the reversible pathways.
- Entropy changes for irreversible processes can be determined by breaking down the change into reversible steps, where the entropy can be precisely defined, with the same initial to final states.
- If the total entropy change, as ΔS_{surr}, is compared to the sum total of the reversible changes, then three possible situations will occur:
  (1) If $\Delta S_{\text{calculated}} < \Delta S_{\text{reversible}}$ then the process will be spontaneous and occur naturally in the direction hypothesized.
  (2) But, if $\Delta S_{\text{calculated}} > \Delta S_{\text{reversible}}$, the process described will be nonspontaneous and will not occur naturally. Energy has to be added to make it happen in the direction indicated.
  (3) If $\Delta S_{\text{calculated}} = \Delta S_{\text{reversible}}$, there will be no net change observed in the system and this describes a reversible or equilibrium situation.

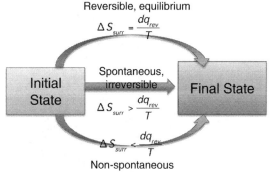

Reversible, equilibrium

$\Delta S_{surr} = \dfrac{dq_{rev}}{T}$

Initial State

Spontaneous, irreversible

$\Delta S_{surr} > \dfrac{dq_{rev}}{T}$

Final State

$\Delta S_{surr} < \dfrac{dq_{rev}}{T}$

Non-spontaneous

$\Delta S_{total} = \Delta S_{surr} + \Delta S_{sys} = 0$ *Equilibrium*

$\Delta S_{total} = \Delta S_{surr} + \Delta S_{sys} > 0$ Spontaneous Change

$\Delta S_{total} = \Delta S_{surr} + \Delta S_{sys} < 0$ Non-spontaneous Change

- The ΔS_{surr} or ΔS_{total} for a reversible adiabatic process is zero, since the ΔS from the volume change is compensated for by the ΔS due to temperature change, so that: $\Delta S_{vol} + \Delta S_{\Delta T} = 0$.
- However, an irreversible, adiabatic process has a non-zero value for ΔS_{surr} or ΔS_{total}.
- Mixing of gases is an example of a spontaneous process where the natural direction is clearly defined. The equation describing ΔS for mixing is given on the map and will always produce a positive value for the ΔS_{surr}.
  - The magnitude of ΔS_{mix} is ONLY dependent on the relative number of particles, the mole fraction of each component gas, as the primary factor.
  - The value of ΔS_{mix} is independent of the volume or pressure at which it occurs.
  - The magnitude of ΔS_{mix} is also not affected by the temperature at which the mixing occurs.

Matter consists of atoms and molecules that have discrete energy states (or microstates), that also have degrees of freedom (such as translational, rotational, and vibrational energies, dependent on the state of matter, size of molecule, bonding, and types of atoms in the molecule). The term W is used as a measure of the number of different "microstates" possible. The Boltzmann equation defines entropy as: $S = k \ln W$. For normal conditions and the number of atoms or molecules, the value of W can be difficult to predict. However, at $T = 0$ K, the value of W, the number of states (arrangements) possible for a perfect crystalline material is only 1, so that $S = k \ln W = k \ln(1) = 0$. But many substances, such as amorphous solids (glasses), cannot form a single uniform state at 0 K and have "residual entropy" or non-zero values for S at 0 K. Other factors that can result in residual entropy are crystalline defects, alterations in structure due to resonance structures or isomers, misalignment of net dipoles, and the like.

EXAMPLE PROBLEMS

3.1 A) Calculate the ΔS_{vap} for the following substances:

Substance	ΔH_{vap} (kJ/mol)	T_{bp} (°C)
CCl_4	30.0	76.7
H_2S	18.7	−60.4
CH_4	8.18	−161.5
H_2O	40.7	100

B) Which substances have similar values for ΔS_{vap}? Considering intermolecular forces characteristic of the substance in the solid and liquid state and the physical change that occurs with vaporization, explain the reason(s) for the similarity.

C) Water appears to be different from the other substances. Using the same criteria as in B), explain why water would have a significantly different value for ΔS_{vap}.

3.2 Considering the substances, toluene and benzene:

A) Toluene, $C_6H_5CH_3$, has a melting point of −95.0°C while that for benzene, C_6H_6, is 5.6°C. If the ΔH_{fus} values are 6.64 kJ/mol and 9.90 kJ/mol, respectively, which substance has the higher ΔS_{fus}? (Prove your choice.)

B) Does the same hold true for the ΔS_{vap} of the two substances, given that, for $C_6H_5CH_3$, $T_{bp} = 111$°C and $\Delta H_{vap} = 361.1$ J/g and where the values for benzene, C_6H_6, are $T_{bp} = 80.1$°C and $\Delta H_{vap} = 394$ J/g?

3.3 The average heat evolved by the oxidation of food by an average adult per hour per kilogram of body weight is 7.20 kJ/kg h. Consider the average heat released to the surroundings by an 80.0 kg adult during a 1-day period.

A) What would be the entropy change in the surroundings, when the surrounding $T = 20$°C?

B) Explain why the ΔS in the surroundings must be calculated from the heat lost by the body and not by one of the defined equations for ΔS.

3.4   Compare the ΔS when 1.0 mol of an ideal monatomic gas is heated from 300 to 500 K when:

A)   The volume is held constant

B)   The pressure is held constant

Furthermore,

C)   Why would there be a difference between the answers for A) and B)?

3.5   When 1.0 mol of $CO_2(g)$ at 298 K is heated to 1000 K, with $P = 1.0$ bar, what would be the ΔS if:

A) $\overline{C}_p$ is assumed to be constant at its 298 K value, 37.14 J/K mol? (p. 82)

B) $\overline{C}_p$ is defined by the function:

$$\overline{C}_p = 18.86 + (0.079377)T + (-6.783 \times 10^{-5})T^2 + (2.443 \times 10^{-8})T^3$$

Furthermore,

C)   Which would be the more trustworthy value to quote? – the value from A) or B)?

3.6   Iodine, $I_2(s)$, has a melting point of 113.8°C and the liquid boils at 185.4°C and has the thermochemical properties given in the box on the right.

Properties for Iodine, I_2	
$\Delta H_{fus} = 15.52$ kJ/mol	$C_p, I_2(s) = 54.44$ J/mol-K
$\Delta H_{vap} = 41.8$ kJ/mol	$C_p, I_2(l) = 80.33$ J/mol-K
	$C_p, I_2(g) = 36.9$ J/mol-K

A)   Given the diagram of heating curve for the solid, describe how to determine the ΔS for each section of the curve.

B)   Calculate ΔS_{total} for the heating of 1 mole of $I_2(s)$ from 30°C to $I_2(g)$ at 200°C.

C)   Which section makes the biggest contribution to ΔS_{total} and what % does it contribute?

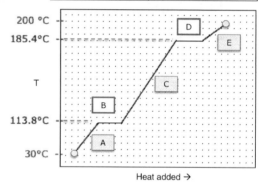

3.7 A)   In an earlier question (2.32) we saw that the C_p for C(s, graphite) changes with T, particularly in the range of 220 to 850 K. Given the C_p for C(s) in the table, what would the ΔS in J/K mol be for C(s) graphite heated from 25° to 600°C?

Heat Capacity C(s), graphite			
	a	b	c
$\overline{C}_p = a + bT + cT^2$	−0.4493	0.03553	−1.3×10⁻⁵
Source: Handbook Chemistry & Physics, 88th edition			

B)   If Fe(s) was also heated from 25° to 600°C, would the ΔS be less than, equal to, or greater than the ΔS for C(s)? Prove your answer.

C)   If C_p for Fe(s) had been taken as a constant of its 298 K value of 24.797 J/K mol, what would be the

Heat Capacity Fe(s)	$\overline{C}_p = A + BT + CT^2 + DT^3$		
A	B	C	D
0.2490	0.02464	−8.91×10⁻⁶	9.66×10⁻⁹
Source: NIST Web book			

value calculated for ΔS for Fe(s) in J/K mol? Comparing this value to that in B), is it necessary to use the function for C_p to determine the ΔS for Fe(s)?

3.8   Compare the ΔS for the dissociation of diatomic molecules $X_2(g) \rightarrow X(g)$ for $Br_2(g)$, $Cl_2(g)$, $I_2(g)$ and $H_2(g)$ into gaseous atoms. What trend(s) do you observe?

3.9   Calculate the ΔS for a sample of Ar(g) at 25°C, 1 atm pressure in a 500-mL container, that is allowed to expand to double its volume and also to be heated to 100°C. Assume the process is a two-step process, where the first step is a reversible, isothermal expansion followed by heating at a constant volume to change the temperature

3.10 When 0.50 mol of an ideal gas is expanded isothermally and reversibly at 298 K from a volume of 10.0 L to 25.0 L,

A)   (a) What is the ΔS_{gas}? (b) How much work was done? (c) What is the ΔS_{surr}? (d) What is the ΔS_{total}?

B)   Suppose the expansion occurs irreversibly, as in allowing the gas to expand isothermally into an evacuated flask ($P=0$) to a total volume of 25.0 L which variables change?
       (a) ΔS_{gas}? (b) The work done? (c) ΔS_{surr}? or (d) ΔS_{total}?

3.11 A normal breath intake results from the lungs expanding to take in air, causing a slight pressure decrease. The typical breath has a volume of 0.50 L taken in every 3 seconds (20 breaths per minute). Suppose the air drawn into the lungs has an initial pressure of 1.0 atm and a $T=20°C$, but inside the lungs the pressure is lowered to 756 torr and the air is heated to 37°C.

A)   What is the ΔS per breath for the air drawn into the lungs?

B)   What is the ΔS_{total} per day for the air drawn into the lungs?

C)   Which is the biggest contributor to the overall ΔS – the temperature change or the pressure change?

3.12 Calculate the ΔS_{mix} when:

A)   10.0 g of $CH_4(g)$ is mixed with 100 g of $C_2H_6(g)$ at 20°C in a 2.0-L vessel.

B)   The vessel size was changed to 20.0-L; would this change the ΔS_{mix}?

MIXING OF GASES

$$\Delta S_{mix} = \Delta S_A + \Delta S_A + \cdots$$

Apply concepts expansion and mixing:

$$\Delta S_{mix} = n_A R \ln\left(\frac{V_A}{V_{total}}\right) + n_B R \ln\left(\frac{V_B}{V_{total}}\right) + \cdots$$

If mixing two gases can simplify as:

$$\chi_A = \frac{n_A}{n_{total}} = \frac{V_A}{V_{mix}} \qquad \chi_B = \frac{n_B}{n_{total}} = \frac{V_B}{V_{mix}}$$

$$\Delta S_{mix} = -R\left[n_A \ln \chi_A + n_B \ln \chi_B\right]$$

3.13 Suppose 10.0 L of $CH_4(g)$ is mixed with 100 L of $C_2H_6(g)$ at 20°C at a constant pressure of 1.0 bar.

A)   Calculate the ΔS_{mix} in J/K.

B)   Compare the ΔS for this mixture to that measured in the previous problem. If similar, why would they be similar?

3.14 Suppose 100 g of $O_2(g)$, 100 g of Ar(g) and 50.0 g $N_2(g)$ are mixed together in a 20.0-L container at 20°C:

A)   What is the total pressure in the container, assuming the gases are acting as ideal gases?

B)   Given the P_{total} calculated in (A), is it reasonable to consider the gases as ideal?

C)   What is the ΔS for the mixture, if the gases are ideal gases?

D)   If the gases were acting as real, not ideal gases, with some interactions occurring, how might this affect the ΔS_{mix}?

3.15 An isolated system consists of a group of four chambers, initially separated by barriers, that each contain a quantity of gas, all at the same P and T, as pictured on the right.

A)   Calculate the $\Delta \overline{S}_{mix}$ when all three barriers are removed simultaneously and the gases allowed to mix at a constant T.

B)   Suppose equal numbers of moles of the gases are mixed in a similar way, but that each section only contains 1.0 mol of each gas. How does the $\Delta \overline{S}_{mix}$ compare between this combination and the $\Delta \overline{S}_{mix}$ value in A)?

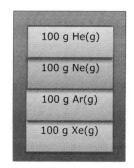

100 g He(g)

100 g Ne(g)

100 g Ar(g)

100 g Xe(g)

3.16 Suppose 100 g of $O_2(g)$ at 50°C was added to 100 g of argon at 20°C under constant pressure conditions in a flexible container that is thermally insulated from the surroundings.

A)   Describe how you could determine the ΔS for mixing.

B)   Show how each calculation could be accomplished and then estimate the answer.

3.17 The freezing of supercooled liquid water would be considered an irreversible process. Calculate the ΔS when 1.0 mol of supercooled liquid water at –10°C and 1 atm pressure turns into ice. (C_p $H_2O(s) = 36.8$ J/K mol)

A)   Construct a diagram showing a series of steps, using known values for entropy, that describe the ΔS for the freezing of supercooled water.

B)   Calculate ΔS for the process.

C) The equation $\Delta S_r^\circ(T_2) = \Delta S_r^\circ(T_1) + \Delta \bar{C}_p \ln\left(\dfrac{T_2}{T_1}\right)$ applies to the calculation of the entropy for a phase change at temperatures different than the "normal" ($P = 1.0$ bar or atm) values. Based on the diagram in A), what is the definition of $\Delta \bar{C}_p$?

3.18 The denaturation of a protein can be viewed as a phase transition, somewhat equivalent to melting in smaller molecules, since only some of the forces are severed. The transition occurs at T_m that is determined using techniques like differential scanning calorimetry (DSC) of the protein that measures the heat capacity of the protein in solution as the temperature changes. The abrupt change in heat capacity signals the phase transition and gives information about the ΔH of the transition (from the area under the peak), as well as the heat capacity before and after the transition (see figure on the right).

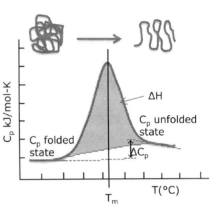

A) Using a stepwise diagram, prove that it is possible to calculate the ΔH and ΔS values for the denaturation, at temperatures other than T_m, defined as:
$$\Delta H_{den}(T_2) = \Delta H_{den}(T_1) + \Delta \bar{C}_{p,den} \Delta T \quad \text{and} \quad \Delta S_{den}(T_2) = \Delta S_{den}(T_1) + \Delta \bar{C}_{p,den} \ln(T_2/T_1)$$
B) A protein denatures with a measured value of 640 kJ/mol at 340 K and $P = 1.0$ bar. The ΔC_p at the transition was 8.37 kJ/mol. What is the:
(a) ΔS for the transition at 340 K?
(b) ΔH and ΔS for the transition at 310 K?

3.19 In a DSC experiment, the T_m of a certain protein was found to be 46°C and the $\Delta H_{den} = 382$ kJ/mol.
A) What would be the ΔS_{den} at 46°C?
B) If there are 122 amino acids in the protein chain, what is the ΔS per amino acid unit in this protein?
C) Compare the value in ΔS for fusion of $H_2O(s)$, $C_6H_6(s)$, $CCl_4(s)$ and $CH_4(s)$ to the value of ΔS per amino acid unit of the protein in B). What does this indicate about the forces holding the protein in its native conformation versus those typical of small molecules in solids?

3.20 A lysozyme denatures by "unfolding" with a $\Delta H = 509$ kJ/mol at 75.5°C and with a ΔC_p at the transition of 6.28 kJ/K mol.
A) What would be the ΔS for the unfolding at 25°C?
B) How much does T affect the value of ΔS for the denaturing or "unfolding"?
C) Based on the units alone, comment on the size of the ΔS for the denaturing process versus other ΔS values for phase transitions, even those as large as vaporization or dissociation calculated in earlier questions.

3.21 Determine the ΔS_{298}° for the following reactions, first by qualitatively judging its sign and then by calculating the value at 298 K from tabled S° values.
A) $6CO_2(g) + 6H_2O(g) \rightarrow C_6H_{12}O_6(s) + 6O_2(g)$
B) $N_2(g) + O_2(g) \rightarrow 2NO(g)$

CHEMICAL REACTION
$aA + bB \rightarrow cC + dD$
$\Delta S_r = \sum S_{298}^\circ, \text{products} - \sum S_{298}^\circ, \text{reactants}$
$\Delta S_r = \left(cS_{298,C}^\circ + dS_{298,D}^\circ\right) - \left(aS_{298,A}^\circ + bS_{298,B}^\circ\right)$
Treat the same way as for ΔH for reaction
If calculating at $T \neq 298$ K:
$\Delta S_r(T_2) = \Delta S_r(T_1) + \Delta \bar{C}_{p,r} \ln\dfrac{T_2}{T_1}$
$\Delta \bar{C}_{p,r} = \left[c\bar{C}_{p,C} + d\bar{C}_{p,D}\right] - \left[a\bar{C}_{p,A} + b\bar{C}_{p,B}\right]$

3.22 For the reaction describing the decomposition of urea:

$$(NH_2)_2CO(s) + H_2O(l) \rightarrow 2NH_3(g) + CO_2(g)$$

- A) What can we expect will be the sign and magnitude of $\Delta S°$ for this reaction? Why can't we predict the sign of $\Delta H°$ as easily as we can that for $\Delta S°$?
- B) Calculate the $\Delta H°$ and $\Delta S°$ for the reaction at 25°C.

3.23 Oxygen gas reacts with glycylglycine, $C_4H_8N_2O_8$, to form urea, $(NH_2)_2CO(s)$, $CO_2(g)$ and $H_2O(l)$.
- A) Write the balanced reaction
- B) Calculate the $\Delta H°$ and $\Delta S°$ for the reaction at 298 K.
- C) Determine the $\Delta H°$ and $\Delta S°$ for the reaction at 330 K.

3.24 A) How do the values for the $\Delta S°$ for the conversions of $C_6H_{12}O_6(s)$ to lactic acid (Reaction I) or to $CO_2(g)$ and ethanol liquid, $C_2H_5OH(l)$ (Reaction II) compare at 298 K?

Reaction (I): $C_6H_{12}O_6(s) \rightarrow 2H_3CH(OH)CO_2H(s)$
Reaction (II): $C_6H_{12}O_6(s) \rightarrow 2CO_2(g) + 2C_2H_5OH(l)$

- B) Determine the $\Delta S°$ values for the reactions at the biological temperature of 37°C.

3.25 If 1.0 mol of an ideal gas expands in a piston arrangement isothermally from a pressure of 10.0 bar to 1.0 bar, what are the values of w, q, ΔU, ΔH, and ΔS when:
- A) The expansion is reversible.
- B) The expansion occurs against an external pressure of zero (into a vacuum).

3.26 One mole of an ideal gas, with $C_v = 5/2\,R$ with $T = 250$ K and $P = 1.00$ bar, undergoes expansion against a constant pressure until its final P is one-half its initial pressure. Calculate q, w, ΔU, ΔH, and ΔS for the transformation:
- A) As an isothermal process
- B) As an adiabatic process

3.27 If 2.5 moles of an ideal gas with $C_v = 3/2\,R$ undergoes the following cyclic process with $T_1 = 450$ K and $P_1 = 2.00$ bar, calculate the q, w, ΔU, ΔH, and ΔS for each step and then for the total process.

| STEP 1: Gas is reversibly and adiabatically expanded until its volume doubles. |
| STEP 2: Gas is reversibly heated at a constant volume until T increases to 450 K. |
| STEP 3: The pressure on the gas is increased in an isothermal reversible compression until $P = 2.00$ bar. |

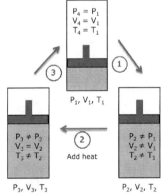

3.28 Two blocks, A and B, of the same metal and the same number of moles, but at different temperatures, are brought into contact and allowed to come to the same final temperature but with no heat gain from or loss to the surroundings.

- A) Prove that $\Delta \bar{S}_{total} = \bar{C}_{p,metal} \ln \left[\dfrac{(T_A + T_B)^2}{4T_A T_B} \right]$

- B) Given the proven equation,
   - (a) Calculate the ΔS for the three temperature combinations if the metal was Fe(s).
   - (b) Which combination would you expect to have the highest ΔS? Give the reason(s) for your expectation.

(c) Why are the values for ΔS different for Sets I and II even though the initial T difference between I and II is the same?

	Set		
	I	**II**	**III**
Block A	100°C	180°C	200°C
Block B	20°C	100°C	600°C

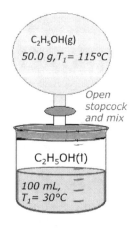

C$_2$H$_5$OH(g)
50.0 g, T_1= 115°C

Open stopcock and mix

C$_2$H$_5$OH(l)

100 mL, T_1= 30°C

C) Would the ΔS_{total} be defined the same way if you have equal moles of different metals for Metals A and B at the different temperatures?

3.29 If 50.0 g of C$_2$H$_5$OH(g) at 115°C are added to 100 mL of C$_2$H$_5$OH(l) at 30°C, in an insulated container, what would be the:
A) Final temperature, in °C, and the state of the mixture? Prove your answer.
B) The ΔS_{total} for the process?

3.30 If six ice cubes at –10°C, with a volume of 15.0 cm^3 each, are added to 400 mL of H$_2$O(l) at 70°C in a thermally insulated container, so that no heat is lost or gained from the surroundings: [$C_{p,m}$ H$_2$O(s) = 37.4 J/K mol]
A) Would all the ice melt?
B) What would be the final temperature of the water mixture?
C) What is ΔS for the ice, ΔS for the liquid (hot) water and ΔS_{total} for the process?

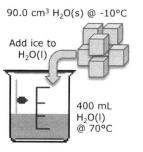

90.0 cm^3 H$_2$O(s) @ -10°C

Add ice to H$_2$O(l)

400 mL H$_2$O(l) @ 70°C

3.31 If a block of copper, weighing 2.0 kg (C_p Cu(s) = 0.385 J/K mol) at 0°C is introduced into an thermally insulated container that contains 1.0 mole of steam, H$_2$O(g), at 100°C and 1 atm, assuming that all the steam is converted to liquid water, what is the:
A) Final temperature of the Cu and the H$_2$O?
B) ΔS total for process?

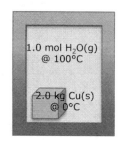

1.0 mol H$_2$O(g) @ 100°C

2.0 kg Cu(s) @ 0°C

3.32 Suppose you have two flasks, both at 20°C, connected by a valve and tubing, containing the gases at the pressures and volumes indicated in the figure below. After the valve is opened:
A) What would be the partial pressure of each gas in the mixture and P_{total}?
B) What would be the ΔS_{mix} in J/K mol for the system?
C) Each gas undergoes an isothermal expansion. Assuming it's reversible,
  (a) Will the work done by each gas be the same or different? If different, what two factors determine the magnitude of the difference?
  (b) Calculate the work done by each gas.

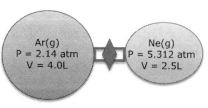

Ar(g)
P = 2.14 atm
V = 4.0L

Ne(g)
P = 5.312 atm
V = 2.5L

3.33 A test tube containing 100 mL of benzene liquid, C$_6$H$_6$(l), was inserted into an ice bath (ice cubes plus water) at 0.20°C. When the test tube was removed from the bath, it was only partially frozen, with 26.7 mL of benzene left as liquid in contact with frozen C$_6$H$_6$. The water bath still had ice present in the water and was still at 0.20°C.

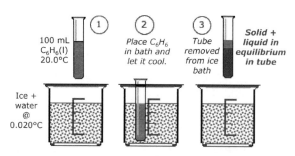

100 mL C$_6$H$_6$(l) 20.0°C

Ice + water @ 0.020°C

① ② ③
Place C$_6$H$_6$ in bath and let it cool.
Tube removed from ice bath

Solid + liquid in equilibrium in tube

A) What was the temperature of the benzene in the tube after it was removed?

B) How would the change in entropy for C_6H_6 be defined? Use the definitions to calculate the value of the change in entropy for C_6H_6.

C) Would the ice bath have also undergone a change in entropy? If yes, calculate the value of ΔS for the ice bath.

D) What can we expect to be true about the sign of the ΔS_{total} for this situation and why? Calculate the value of ΔS_{total}. Does it match your expectation?

E) What if the tube of C_6H_6, once removed, had been completely frozen and reached the bath temperature? How would this have changed the calculation(s) for:
   (a) ΔS in B)?
   (b) ΔS in C)?
   (c) ΔS in D)?
   (d) Under these circumstances, calculate the changed values in B), C), and D) and then the new value of ΔS_{total} when the C_6H_6 completely freezes.

KEY POINTS – REAL GASES AND ENTROPY

As discussed in Part 1, if a gas acts a "real gas", its molar volume can be affected. This should have an impact on changes in entropy, due to isothermal volume changes. The following two examples illustrate the potential impact on entropy.

3.34 A) Prove that, for an isothermal reversible expansion or a contraction of a van der Waals gas, the volume dependence of entropy can be defined by the equation below, if the second term involving "a" can be ignored.

$$\Delta S_{VdW} = R\ln\left[\frac{\bar{V}_2 - b}{\bar{V}_1 - b}\right]$$

B) Determine the ΔS for $CO_2(g)$ and $CCl_4(g)$ for an isothermal reversible volume change from 0.50 to 10 L/mol as an (a) ideal gas and then (b) as a van der Waals gas, using the equation in A).

3.35 For an isothermal reversible volume change for a gas:

A) Prove that, for a virial gas, obeying the virial equation, $P_{virial} = \dfrac{RT}{V_m}\left[1 + \dfrac{B}{V_m}\right]$ produces the entropy: $\Delta S_{Virial} = R\ln\left[\dfrac{V_{m2}}{V_{m1}}\right] + BR\left[\dfrac{1}{V_{m1}} - \dfrac{1}{V_{m2}}\right]$ for the volume change.

B) Determine the ΔS for $CO_2(g)$ and $CCl_4(g)$ for volume change from 0.50 L/mol to 10 L/mol, at 340 K and as (a) an ideal gas and then as (b) a gas obeying the equation derived above in A), where B $CO_2(g) = -108$ cm^3/mol and B $CCl_4(g) = -1171$ cm^3/mol (from D. R. Lide (ed.) *CRC Handbook of Chemistry and Physics*, 88th edition, CRC Press).

C) Compare the results for $CO_2(g)$ and $CCl_4(g)$:
   (a) As a virial gas compared to the value of an ideal gas.
   (b) As a van der Waals gas (using the first term only) in Problem 3.34 compared to those when they act as a virial gas. If they are different, what factor(s) could explain the difference?
   (c) Why did we need to know the temperature for the virial gas, but not for the van der Waals gas?

Free Energy (ΔG), Helmholtz Energy (ΔA), and Phase Equilibrium

4

KEY POINTS – GIBBS AND HELMHOLTZ FREE ENERGY

The **Clausius inequality** $TdS \geq dq- = dU - dw$ gives the conditions for spontaneous change as $-dU + dw + TdS \geq 0$, where the sum of the three terms can only equal zero for a reversible process.

- Work then needs to be defined by two terms: $dw = -PdV + dw_{\text{non-}PV}$.
- Non-PV work is not due to expansion or contraction and is work that can be extracted from the process such as electrical work, elastic work of stretching, or the unfolding of a protein.

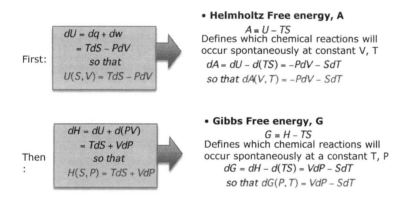

First:
$$dU = dq + dw$$
$$= TdS - PdV$$
so that
$$U(S,V) = TdS - PdV$$

• **Helmholtz Free energy, A**
$$A = U - TS$$
Defines which chemical reactions will occur spontaneously at constant V, T
$$dA = dU - d(TS) = -PdV - SdT$$
so that $dA(V,T) = -PdV - SdT$

Then:
$$dH = dU + d(PV)$$
$$= TdS + VdP$$
so that
$$H(S,P) = TdS + VdP$$

• **Gibbs Free energy, G**
$$G = H - TS$$
Defines which chemical reactions will occur spontaneously at a constant T, P
$$dG = dH - d(TS) = VdP - SdT$$
so that $dG(P,T) = VdP - SdT$

Once the second law of thermodynamics was introduced, dU and dH could also be redefined, as shown on the right. Two kinds of "free energy" (non-PV work) are then defined, the sign of which will also determine whether the process is spontaneous or non-spontaneous.

Because most systems of interest in chemistry and biochemistry are open systems, and P is constant, the Gibbs free energy, G or ΔG, is the most useful and summarizes the consequences of both the first law (with dH) and second law (with dS) on chemical processes.

By using Gibbs free energy, G (or Helmholtz energy, A), it is no longer necessary to consider dS_{surr} for a described change in the system to define the direction of natural change. Knowledge of dG or dA for the system alone will be sufficient to predict the direction of natural change. As long as dG (or dA) decreases for the process, meaning dG or dA is negative, then $dS_{\text{surr}} = (dq / T) \geq 0$ and the change will be spontaneous. On the macroscopic scale, we can say that:

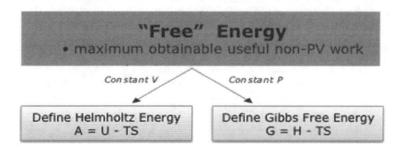

- $\Delta G < 0$ ($\Delta A < 0$) is a spontaneous (irreversible) process – non-PV work can be extracted
- $\Delta G = 0$ ($\Delta A = 0$) is a reversible (equilibrium) process
- $\Delta G > 0$ ($\Delta A > 0$) is a non-spontaneous process, requiring energy input to occur, since it is opposite to the natural direction.

Gibbs free energy, ΔG, is the maximum amount of non-PV work that can be extracted from the reaction at constant P and T, while ΔA is the maximum amount of non-PV work obtained from constant V isothermal processes.

For isothermal chemical change, $dG - dH - TdS$ becomes $\Delta G = \Delta H - T\Delta S$ and is a very useful equation in both chemistry and biochemistry. Considering ΔG for chemical changes or changes in state at constant P, T:

Isothermal Change

$$dG = dH - TdS \text{ leads to } \Delta G = \Delta H - T\Delta S$$

Process with T Change

$$dG = dH - TdS \text{ leads to: } dG = VdP - SdT$$

CHEMICAL REACTION

$$aA + bB \rightarrow cC + dD$$

$$\Delta G_r^\circ = \sum \Delta G_{f,\text{products}}^\circ + \sum \Delta G_{f,\text{reactants}}^\circ$$

$$\Delta G_f^\circ = \left(c\Delta G_{f,D}^\circ + d\Delta G_{f,D}^\circ\right) - \left(a\Delta G_{f,A}^\circ + b\Delta G_{f,B}^\circ\right)$$

ΔG° can change sign with T only when ΔH° and ΔS° are the same sign

$$T_{\text{sign change}} = \frac{\Delta H_r^\circ}{\Delta S_r^\circ}$$

- Chemical transformations will always be spontaneous when $\Delta H(-)$ and $\Delta S(+)$.
- Chemical transformations will always be non-spontaneous when $\Delta H(+)$ and $\Delta S(-)$.
- When the signs of ΔH and ΔS are the same, the sign of ΔG will depend on T and there will be a T at which $\Delta H = T\Delta S$ that will represent the "changeover" from spontaneous to non-spontaneous for these chemical reactions.

(These same guidelines apply to $\Delta A = \Delta U - T\Delta S$, when the changes occur at fixed V and T.)

The pressure and temperature dependence of the Gibbs Free energy is obtained by considering that dG must be defined by $dG(P,T) = [\delta G/\delta T]_P \, dT + [\delta G/\delta P]_T \, dP$ and also by $dG(P,T) = VdP - SdT$ so that:

- $$\left[\frac{dG}{dT}\right]_P = -S \text{ and } \left[\frac{dG}{dP}\right]_T = \bar{V} \quad \left[\frac{\delta G}{\delta P}\right]_T dP = \bar{V} \Rightarrow \frac{dG}{dP} = \bar{V}$$

- So, G decreases (in magnitude) when T increases since entropy must always increase with T (see graph from map).
- Whereas G increases when P increases (see graph from map).

The **temperature dependence of G or ΔG** is the same for all states as well as for chemical changes, summarized as the Gibbs–Helmholtz equation:

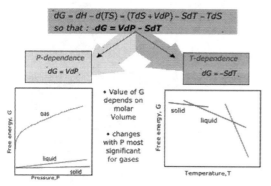

$dG = dH - d(TS) - (TdS + VdP) - SdT - TdS$
so that : $dG = VdP - SdT$

P-dependence
$dG = VdP$

T-dependence
$dG = -SdT$

- Value of G depends on molar Volume
- changes with P most significant for gases

$$\frac{G(T_2)}{T_2} = \frac{G(T_1)}{T_1} + H_{(T_1)}\left[\frac{1}{T_2} - \frac{1}{T_1}\right] \text{ or } \frac{\Delta G(T_2)}{T_2} = \frac{\Delta G(T_1)}{T_1} + \Delta H_{(T_1)}\left[\frac{1}{T_2} - \frac{1}{T_1}\right]$$

The **pressure dependence of G or ΔG** for solids and liquids is different than for gases, as shown on the map.

- Solids or liquids: $G(P_2) = G(P_1) + V\Delta P$
- Gases: $G(P_2) = G(P_1) + RT \ln \frac{P_2}{P_1}$

Besides defining how dG changes in response to changes in P or T, these equations also help to define why changes in state occur at single temperatures, to identify which states will be most stable at a certain P or T, and to determine what is required for equilibrium between two states (phases) of a substance.

For pure substances, the molar free energy, $\bar{G}_i$, can be defined as the **chemical potential, μ**. The chemical potential defines the criteria for equilibrium to exist between two phases or states. A **phase** is when a substance has a uniform chemical composition and physical structure for a range of P and T values. A phase diagram shows the thermodynamically stable phase at any P and T values and when two phases (such as solid and liquid) are in equilibrium with each other.

- *The phase with the lowest μ under given conditions is the most stable state, based on thermodynamics.*
- *The Clapeyron and Clausius–Clapeyron equations are defined for the equilibrium of two phases, as shown on the map.*

Equilibrium is the coexistence of two states where the rate of the forward process equals the rate of the reverse process, so that no net change is observable in the bulk properties or appearance of the substance.

- *When two phases (α, β) are in equilibrium: $\mu_\alpha = \mu_\beta$, so that $\Delta\mu = 0$.*
- *If $\Delta\mu < 0$, a spontaneous transfer of matter will occur between the two phases.*

To become an intensive property (independent of the amount of matter), the free energy, dG, must be described as a function of n as well as P and T, so that: $dG = VdP - SdT + \sum_i \mu_i dn_i$

The non-expansion work can arise from changing the composition of a system that is not at equilibrium, but at a constant T and P, and is defined as: $dG = \sum_i \mu_i dn_i$.

Both pure solids and liquids have vapor pressures, so that the chemical potential is then defined as: $\mu_i = \mu_i^\circ + RT\ln(P/P^\circ)$ where $P^\circ = 1$ atm (or 1.0 bar) and P is equal to the vapor pressure in atm (or bars).

The appropriate ΔH must be applied to the change in state or transition between the two phases:

Solid → liquid	$\Delta\bar{H}_{fusion} = \Delta\bar{H}_{fus}$	Normal T°_{fp} when $\mu_{solid} = \mu_{liquid}$ at 1.0 atm (1.0 bar)
Solid → gas	$\Delta\bar{H}_{sublimation} = \Delta\bar{H}_{sub}$	Normal T°_{sub} when $\mu_{solid} = \mu_{gas}$ at 1.0 atm (1.0 bar)
Liquid → gas	$\Delta\bar{H}_{vaporization} = \Delta\bar{H}_{vap}$	Normal T°_{bp} when P_{vap} equals 1.0 atm (1.0 bar)

The temperature dependence of phase transition ΔH_{tr} for $\alpha \to \beta$ can be defined as:

$$\Delta H^\circ_{tr,\alpha\to\beta}(T_2) = \Delta H^\circ_{tr,\alpha\to\beta}(T_1) + \left[\Delta\bar{C}_{p,\alpha\to\beta}\right]\Delta T$$

EXAMPLE PROBLEMS

4.1 Given the graph of molar free energy for substance X at some constant P shown on the right,
- A) Which of the temperatures ($T_1 \to T_4$)
  - (a) Represents single phases and which, phase transitions?
  - (b) For each single phase, what is the thermodynamically stable state?
  - (c) For the phase transition(s), what two states are in equilibrium?
- B) Explain why substance X sublimes before it melts.

4.2 For the statements below, CIRCLE the word or words inside the parentheses that serve to make a correct statement. Each statement has at least one but may have more than one correct choice.
- A) Given a constant P, a chemical change will always be spontaneous when the entropy of the change (is positive, remains constant, is negative) and the enthalpy (is positive, remains constant, is negative) during the process.

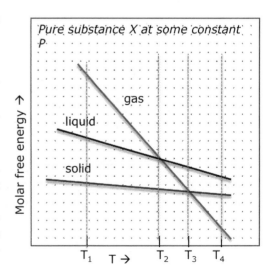

B)   When P is increased on a pure substance, at constant T, the molar free energy, $\bar{G}$ (increases, remains constant, decreases).

C)   When a sample of liquid is converted reversibly to its vapor at its normal boiling point (q, w, ΔP, ΔV, ΔT, ΔU, ΔH, ΔS, ΔG, or none of these) is equal to zero for the system.

4.3   If 2.0 moles of an ideal gas are expanded isothermally from 25.0 L to 125 L at 340 K, calculate the ΔS, ΔG, and ΔA for the change.

4.4   For each of the following processes, state which (if any) of the variables, ΔU, ΔH, ΔS, or ΔG, is equal to zero:

A)   Isothermal reversible expansion of an ideal gas.

B)   Adiabatic reversible expansion of a non-ideal gas.

C)   Vaporization of liquid water at 80°C and 1 bar pressure.

D)   Vaporization of liquid water at 100°C and 1 bar pressure.

E)   Reaction between H_2 and O_2 in a thermally insulated bomb calorimeter.

F)   Reaction between H_2SO_4 and NaOH in dilute aqueous solution at constant temperature and pressure.

4.5   Fuel values are measured as kJ/g for the combustion of the fuel.

A)   For the combustion of $C_6H_6(l)$ (given that $H_2O(l)$ is formed as a product):

(a)   Calculate the maximum of non-PV work that can be extracted per gram from its ΔH and ΔS values.

(b)   Will the reaction always be spontaneous? How can the combustion be stopped?

B)   Calculate the maximum of non-PV work that can be extracted per gram from combustion of $H_2(g)$. Which is the better fuel on a per gram basis?

4.6   The ΔH of formation of 1 mole of NOCl(g) from the gaseous elements at 25°C is 51.9 kJ.

A)   What is the ΔS_f and ΔG_f at 25°C for the reaction?

B)   Assuming all gases act as an ideal gas, what is the ΔU and ΔA at 25°C for the formation reaction?

C)   Given the values of ΔG and ΔA, will the reaction be: always spontaneous, always non-spontaneous or spontaneous only at some T values? Prove your choice.

4.7   For the reaction of cyclopentene, C_5H_8, with $I_2(s)$:

$$C_5H_8(l) + I_2(s) \rightarrow C_5H_6(l) + 2HI(g)$$

	$C_5H_8(l)$
ΔH_f°	58.2 kJ/mol
S°	291.3 J/mol-K
	$C_5H_6(l)$
ΔH_f°	134.3 kJ/mol
S°	274.3 J/mol-K

A)   What is the ΔH°, ΔS°, and ΔG°, for the reaction at 25°C?

B)   What is the ΔU° and ΔA° for the reaction at 25°C?

C)   Is there a temperature at which the sign of ΔG will change? If so, calculate the T and determine if ΔA will also change sign at the same temperature.

4.8   When nitroglycerin decomposes, the chemical reaction that occurs can be:

$$4C_3H_5N_3O_9(l) \rightarrow 12CO(g) + 10H_2O(g) + 6N_2(g) + 7O_2(g)$$

A)   Calculate the ΔH°, ΔS°, and ΔG° for this reaction, given ΔH_f° for $C_3H_5N_3O_9(l) = -370$ kJ/mol and its $S^\circ = 545$ J/K mol.

B)   What can you say about the spontaneity of this reaction as the T changes?

C)   Calculate the ΔU° and ΔA° for the reaction

D)   By comparing the magnitudes, how much more non-PV work can be extracted as ΔA° than ΔG° for this reaction, on a percentage basis?

4.9   Tin (Sn) has two major crystalline forms, gray tin and white tin, with the properties described in the box on the right. The beta form is the familiar common form of tin with the metallic luster, and it is the most thermodynamically stable form above 13.2°C. Below 13.2°C, pure tin can spontaneously transform itself into the alpha form, as a powdery dust. This transformation is called "zinc pest" and can only be avoided by alloying the tin with small amounts of other metals.

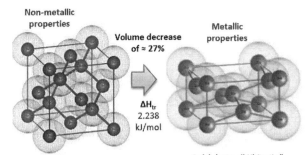

Non-metallic properties

Metallic properties

Volume decrease of ≈ 27%

ΔH_{tr} 2.238 kJ/mol

Sn(s), alpha "Gray tin" (diamond-like structure) Density = 5.77 g/cm³ (13°C)

Sn(s), beta "White tin" (body centered tetrahedral) Density = 7.31 g/cm³ (15°C)

A)   What is the ΔS for the $\alpha \to \beta$ conversion of the gray to the white tin form at 13.2°C?

B)   Based on the signs of ΔH and ΔS, will changes in T affect the spontaneity of the $\alpha \to \beta$ conversion?

C)   Considering the conversion in terms of $\mu_\beta - \mu_\alpha$, prove which form will be more stable at 0°C.

D)   If you were to increase the pressure on the beta form by 10 atm, prove which form would then be the more stable.

4.10   Suppose you have 2.0 moles of the following substances at 20°C and 2.0 atm and the pressure is changed to 200 atm, assuming ideal gas behavior for gases.

A)   Calculate the change in the free energy if the substance is:
   (a)   methane, CH_4, gas
   (b)   ethane, C_2H_6, gas
   (c)   benzene, C_6H_6, liquid

B)   Which remain the same and why?

C)   For those that are different, why are they different?

4.11   Creatures that live in the deep ocean experience very high external pressures due to the weight of water, combined with gravity. Therefore, for every 33 feet of depth (10.06 m), the pressure increases by 14.5 psi over atmospheric pressure at sea level. Suppose you brought a sea creature, such as a squid, up to the surface from a depth of 2.0 km:

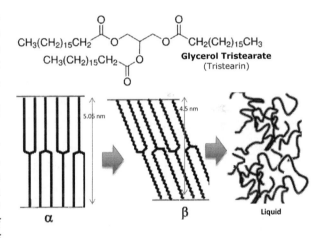

$CH_3(CH_2)_{15}CH_2$ $CH_3(CH_2)_{15}CH_2$ $CH_2(CH_2)_{15}CH_3$

Glycerol Tristearate (Tristearin)

5.05 nm   4.5 nm

α   β   Liquid

A)   What would the ΔP be, in atm, that the squid would experience? [1 psi = 51.7 torr]

B)   Glycerol tristearate, $C_{57}H_{111}O_6$ [MW 891.5], is a common fat in all living creatures that has two solid phases, alpha and beta. The alpha form converts to the beta form at 54°C with $\Delta H_{tr} = 145$ kJ/mol and the beta form then melts at 74°C ($P = 1.0$ atm) with a $\Delta H_{fus} = 203.3$ kJ/mol [see figure]. The solid (beta) form has a density of 0.950 g/mL and the liquid form, 0.856 g/mL at 90°C. If the glycerol tristearate in the squid undergoes the pressure change in A):
   (a)   What would be the ΔG for the fat?
   (b)   What would be the change in melting point of the beta form because of the change in pressure? Would the transition temperature from alpha to beta show the same change?

C)   Determine the $\Delta S°$ of the $\alpha \to \beta$ transition at 54°C and for melting at 74°C of glycerol tristearate, $C_{57}H_{111}O_6$.
   (a)   How does the $\Delta S°$ for the transition compare in size to that of the melting of the fat? What does that say about the structural changes at the two transitions?
   (b)   Compare the $\Delta S°$ for the $\alpha \to \beta$ transition for the fat to that of Sn in Problem 4.8. What does that say about the structural changes at the two transitions?

4.12 The unfolding of a protein can be considered to be a change from phase A to phase B at equilibrium temperature T_m and equilibrium pressure of 1.0 atm. At these equilibrium conditions, the heat absorbed per mol of material undergoing this transition is q_{tr} and there is a molar volume change of ΔV_m and ΔC_p, as discussed in Part 2. A form of lysozyme unfolds with a $q_{tr} = 544$ kJ/mol with $\Delta C_p = 6.28$ kJ/mol K at $T_{tr} = 71.4°C$, and $P_{tr} = 1$ atm.

A) Calculate ΔH_{den}, ΔS_{den}, and ΔG_{den} for the unfolding of the protein at 71.4°C.

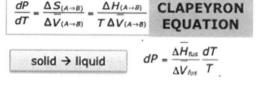

$$\frac{dP}{dT} = \frac{\Delta \overline{S}_{(A \to B)}}{\Delta V_{(A \to B)}} = \frac{\Delta H_{(A \to B)}}{T \Delta V_{(A \to B)}}$$ **CLAPEYRON EQUATION**

B) Calculate ΔH_{den}, ΔS_{den}, and ΔG_{den} for converting one mole of the system from Phase A to Phase B at 37°C and 1.0 atm, using the terms given above, taking into account ΔC_p.

solid → liquid $dP = \dfrac{\Delta H_{fus}}{\Delta V_{fus}} \dfrac{dT}{T}$

C) If, instead of using the approach in B), you assumed that you could calculate the new ΔG at 37°C by just changing the T in $\Delta G = \Delta H - T\Delta S$, how big an error would you have made? Why shouldn't we ignore the ΔC_p term in protein folding/unfolding transitions, if it is known?

D) If the ΔV_m for the protein unfolding is 4.58 cm³/mol, how would a pressure increase of 10 atm affect the transition temperature for the unfolding.

4.13 Solid sulfur is made up of rings of 8 sulfur atoms, in either a rhombic (α) form or a monoclinic form (β), which undergo a reversible transition at 95.5°C. Given the data on the figure:

Rhombic crystal S₈(s)   **Monoclinic crystal S₈(s)**

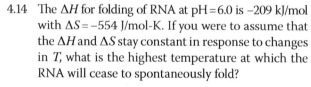

$\alpha \longrightarrow \beta$

$T_{tr} = 95.5°C$

A) Calculate the $\Delta S°$ for the transition from the tabled $\Delta H_{298}°$ and $S_{298}°$ values for the two forms.

B) Calculate the values of ΔH_{tr} and ΔS_{tr} at the transition at 95.5°C.

$C_p = 22.64$ J/mol-K
Density = 2.07 g/cm³

$C_p = 23.6$ J/mol-K
Density = 1.96 g/cm³

C) What pressure would need to be applied to solid sulfur to make the rhombic form stable at 100°C?

4.14 The ΔH for folding of RNA at pH = 6.0 is –209 kJ/mol with $\Delta S = -554$ J/mol-K. If you were to assume that the ΔH and ΔS stay constant in response to changes in T, what is the highest temperature at which the RNA will cease to spontaneously fold?

4.15 Given the data for Hg in the table on the right,

A) What is the melting point of Hg(s) under a pressure of 100 atm?

B) What is the % change in the melting point?

Properties of Hg	
Solid	**Liquid**
T_{fp} –38.9°C	T_{bp} 356°C
ΔH_{fus} 2.292 kJ/mol	ΔH_{vap} 59.30 kJ/mol
Density at –38.9°C	
14.193 g/mL	13.690 g/mL

4.16 It has been suggested that surface melting of ice under a speed skater's skate is responsible for their fast speeds. To test the hypothesis, if the density of ice is 920 kg/m³ and that of $H_2O(l)$ is 997 kg/m³ at 0°C and $\Delta H_{fus} = 6010$ J/mol:

A) What pressure is required to melt ice at –5°C (rink ice temperature)?

B) If skate length is 25.0 cm, its sharpened edge has a width of 0.0250 cm and the skater weighs 85.0 kg, how much pressure can the skater put on the skate edge? Is it enough to melt the ice?

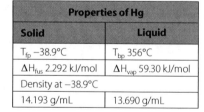

$$P = \frac{force}{area} = \frac{N}{area}$$

$$1\,Pa = \frac{N}{m^2} = \frac{kg}{m\text{-}s^2}$$

$$g = 9.807 \frac{m}{s^2}$$

4.17 At 100°C and 1.0 atm, the V_m of $H_2O(g)$ is 30.20 L/mol and that of $H_2O(l)$ is 0.0188 L/mol.
  A) What would be the change in boiling point per atm of pressure, dT/dP, assuming the molar volumes stayed constant?
  B) Would the boiling point increase or decrease with an increase in pressure?
  C) Why must we still use T in degrees K instead of being able to substitute T in °C?
  D) Compare the result for the change in boiling point in this problem to the change in melting point calculated in Problem 4.15. What factor(s) would be the primary reason for the difference?

4.18 If you were a hiker boiling water at the top of a mountain peak at a height of 5500 feet above sea level, where the atmospheric pressure is 380 torr.
  A) Would you expect that the water temperature when the water boils is below, higher than or equal to 100°C? Briefly, give the reason(s) for your choice.
  B) Which equation would be used to determine the boiling point temperature, the Clapeyron or the Clausius–Clapeyron equation? Briefly, give the reason(s) for your choice.
  C) Calculate the boiling point temperature of the water at this elevation.

4.19 A newer cooking method involves using "instant pots", which can function as pressure cookers (or slow cookers). Pressure cookers are useful because the food can be cooked to doneness in a very short period of time, compared to cooking on the stovetop or in the oven. A typical pressure setting for the pot is "10 psi" which is the pressure added to atmospheric pressure for the cooking process. Given that you are cooking at the boiling point of $H_2O(l)$, and you have the information given in Problem 4.16, how high a temperature are you cooking at when you set the additional pressure to 10 psi? (1.0 psi = 51.7 torr)

4.20 The normal boiling point of liquid benzene is 80.2°C and its vapor pressure is 0.10 bar at 20°C. Calculate the value of:
  A) ΔH_{vap}
  B) ΔS_{vap} at the normal T_{bp} in J/mol K

4.21 The ΔH_{vap} of $H_2O(l)$ at 100°C is 40.66 kJ/mol when $P = 1.0$ atm.
  A) Calculate the w, ΔU, ΔG, and ΔS values for the vaporization 1.0 mole $H_2O(l)$ at 1.0 atm.
  B) What would the ΔG value be at 10 atm and 100°C? Would the vaporization be spontaneous at the higher pressure?
  C) What would be true about the spontaneity of the vaporization if the pressure were lowered, say, to 0.500 atm at 100°C?

| liquid → gas | OR | solid → gas |

$$\Delta V_{tr} = V_{gas} - V_{liquid\,(solid)} \approx V_{gas}$$

$$V_{gas}dP = \frac{\Delta H_{tr}}{T_{tr}}dT$$
• Assume ideal gas
• Integrate to get changes in vapor pressure (P) with T

$$\ln P_2 = \ln P_1 - \frac{\Delta H_{tr}}{R}\left[\frac{1}{T_2} - \frac{1}{T_1}\right]$$

CLAUSIUS-CLAPEYRON EQUATION

4.22 Consider the reaction that converts pyruvic acid (CH_3COCO_2H) into acetaldehyde (CH_3CHO) and gaseous CO_2, which is catalyzed in aqueous solution by the enzyme pyruvate decarboxylase. Assume ideal gas behavior for the CO_2.
  A) Calculate ΔG_{298} for this reaction.
  B) What does Le Chatelier's principle predict to be the effect of increasing the pressure on the reaction to 100 atm?
  C) Calculate ΔG for this reaction at 298 K and 100 atm. State any important assumptions needed in addition to ideal gas behavior.

4.23 A butane lighter contains liquid butane, $C_4H_{10}(l)$, that has a low boiling temperature and it's the vapor that actually burns and provides the flame, once ignited. The normal boiling point of butane is –1.0°C and its $\Delta H_{vap} = 23.8$ kJ/mol. What is the:
  A) Vapor pressure in a lighter at 27°C if the boiling point of butane is –1.0°C at 1.0 atm?
  B) ΔG_{vap} at the vapor pressure calculated in A) and how does this explain why we see liquid butane in the lighter at 27°C rather than just vapor?

4.24 A tissue specimen may be preserved by freeze-drying it, but the T cannot be lower than –10.5°C, as the sample would decompose at this temperature or lower. If the vapor pressure of water is 4.583 mmHg at 0°C, what pressure would be needed for water to sublime from the sample at –10.5°C? (Assume $\Delta H_{sub} = \Delta H_{vap} + \Delta H_{fus}$)

4.25 Suppose there are three small bottles that have lost their labels in an organic chemistry lab. You have access to a device that will let you measure the vapor pressures of the liquid at 25°C very easily. The likely three liquids are given below with their ΔH_{vap} and normal T_{bp}. Could you identify which liquid is in each bottle just by measuring the vapor pressure alone? Prove your approach.

	Ethanol	Acetone	n-Hexane
Normal T_{bp} (K)	351.65	329.65	490.9
ΔH_{vap} (kJ/mol)	39.32	30.25	28.85

4.26 Liquid A is in equilibrium with its vapor at 300 K with a vapor pressure of 40 torr and the ΔH_{vap} for A(l) is 8.00 kJ/mol. The C_p of Liquid A is 67.0 J/K mol and that for the gas of Liquid A is 35.0 J/K mol.

A)   What would be the equilibrium vapor pressure at 350 K, assuming ΔH_{vap} is constant?

B)   What would be the equilibrium vapor pressure at 350 K, if you calculated the ΔH_{vap} at 350 K using the C_p values?

C)   Given the difference between A) and B), should the ΔC_p term be included when calculating P_2? If so, what factor(s) influence whether or not it has a significant effect?

4.27 The Clausius–Clapeyron equation can be viewed as a linear equation, $y = mx + b$, where x, y are the dependent variables whereas m and b are constants or groups of constants.

A)   What experimental values do the y and x correspond to?

B)   What values of interest could you get by plotting y versus x from:

(a)   The slope of the line?

(b)   The y-intercept?

(c)   What units must the vapor pressure and temperature have, before plotting, to obtain the correct units on these values?

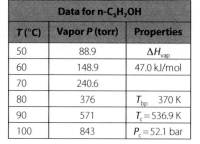

Data for n-C₃H₇OH		
T (°C)	**Vapor P (torr)**	**Properties**
50	88.9	ΔH_{vap}
60	148.9	47.0 kJ/mol
70	240.6	
80	376	T_{bp} 370 K
90	571	$T_c = 536.9$ K
100	843	$P_c = 52.1$ bar

C)   The vapor pressure for n-propyl alcohol, at various temperatures, was plotted, as shown on the right. Describe how you could determine the ΔH_{vap} of the alcohol and its normal boiling point from the graph.

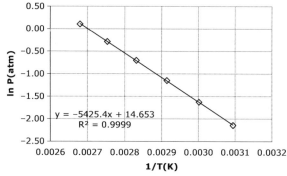

D)   Calculate the values of ΔH_{vap} and its normal boiling point, $T_{bp}°$, using the approach described in C) and compare the results to the tabled values in the box.

E)   Given the P_c for the liquid (given in the box), calculate T_c from the equation of the line. Compare the calculated value to the tabled value. If there is a significant difference, which assumption(s) made in the original derivation of the Clausius–Clapeyron equation might no longer be correct?

4.28 As we have seen in earlier problems, the ΔC_p may be significant and may affect ΔH_{tr} for the phase change. CS_2(l) has a vapor pressure of 40 and 100 torr at 250.65 K and 268.05 K, respectively. What is the ΔH_{vap} and normal boiling point when:

A) ΔC_p is ignored?

B) ΔC_p is included, given that C_p(l) is 75.7 J/mol K and C_p(g) is 45.4 J/mol K for CS_2.

C)   Given that the literature value for the boiling point of CS_2 is 46.3°C, is the adjustment with the ΔC_p term warranted?

KEY POINTS – THE CHEMICAL POTENTIAL AND PHASE DIAGRAMS

Phase diagrams are summaries of the existence of the major phases – solid, liquid, and gas – for any pure substance. The diagram can be constructed by determining the P and T combinations, that produce equilibrium between a minimum of two of the three phases, using the Clapeyron and Clausius–Clapeyron equations, derived from $\mu_\alpha = \mu_\beta$.

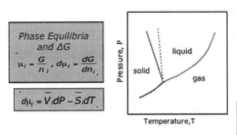

- Lowest chemical potential most likely state at any P,T
- lines represent equilibrium between two states

Phase Rule :
$F = C - P + 2$

Within the phase diagram the "**phase rule**" applies: $F = C - P + 2$ where:

F = degrees of freedom, the number of intensive variables (P, T, χ) that can be changed independently and still maintain all phases.

C = the number of components needed to define the composition of the phase and

P = the number of phases that exist at that point, as defined by phase boundaries.

The **triple point** is the single point on the phase diagram where all three phases co-exist and is defined by finding T at which the vapor pressure of the solid equals the vapor pressure of the liquid:

$$T_{\text{triple point}} = \frac{\Delta \bar{H}_{\text{vap}} - \Delta \bar{H}_{\text{sub}}}{R\ln\left(\dfrac{P_{\text{vap,liquid}}}{P_{\text{vap,solid}}}\right) - \dfrac{\Delta \bar{H}_{\text{sub}}}{T_{\text{solid}}} + \dfrac{\Delta \bar{H}_{\text{vap}}}{T_{\text{liquid}}}}$$

Given two sets of P and T:

$P_{\text{vap,solid}}$ at T_{solid}
$P_{\text{vap,liquid}}$ at T_{liquid}

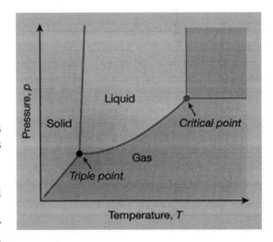

- A single point marks the end of the liquid region, the **critical point,** as discussed in Part 1. In the "supercritical" region, the liquid and gas states have the same density.
- The liquid cannot exist above T_c, P_c.
- The slope of the solid–liquid line is set by the Clapeyron equation and the sign of ΔV_m.
- Many phases may exist in the solid state for some substances, and their equilibrium lines are also determined by the Clapeyron or Clausius–Clapeyron equations.

EXAMPLE PROBLEMS

4.29 We can estimate the triple point from determining the temperature at which the solid and liquid have the same vapor pressure, using the Clausius–Clapeyron equation.

$$T_{\text{triple pt}} = \frac{\Delta \bar{H}_{\text{vap}} - \Delta \bar{H}_{\text{sub}}}{R\ln\left(\dfrac{P_{\text{vap,liquid}}}{P_{\text{vap,solid}}}\right) - \dfrac{\Delta \bar{H}_{\text{sub}}}{T_{\text{solid}}} + \dfrac{\Delta \bar{H}_{\text{vap}}}{T_{\text{liquid}}}}$$

A) Derive the triple point equation given above.
B) Naphthalene, $C_{10}H_8$ (used in mothballs), melts at 353.2 K and has the vapor pressures for the solid and the liquid phases given in the table below. From these data, determine the:
  (a) ΔH_{vap} and ΔH_{sub} for naphthalene
  (b) ΔH_{fus}, from the ΔH_{sub} and ΔH_{vap}
  (c) The normal boiling point of $C_{10}H_8(l)$.
  (d) The triple point for naphthalene.

T (K)	Phase	Vapor Pressure
333.4	solid	0.2448 kPa
349.9	solid	0.7893 kPa
356.5	liquid	1.1744 kPa
464.85	liquid	53.774 kPa
Source: Dortmund Data Bank		

4.30 For the phase diagram and heating curve for substance X on the right:
  A)   Indicate at which lettered point(s) in Diagram A would it be true that:
     (1)   An equilibrium exists between:
        (a) Deposition and sublimation
        (b) Vaporization and condensation
        (c) Crystallization and melting
     (2)   X exists as single phase where X is a:

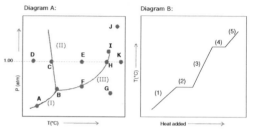

        (a) Solid   (b) liquid   (c) gas
     (3)   There would be three phases present
     (4)   It represents the beginning of the supercritical fluid region
     (5)   It would change position if the density of the X(l) was less than X(s)
  B)   State which points on the phase Diagram A will fall in regions (1)–(5) on the heating curve Diagram B when the $P = 1.0$ atm.
  C)   Indicate the state(s) of X in the sequence $G \rightarrow H \rightarrow I \rightarrow J$.

4.31 Given the phase diagram for CO_2 on the right, within what range of values of P, T could the following conditions be met:

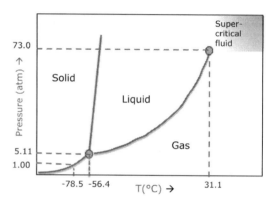

  A)   As T is increased, the solid is first converted to a liquid and then to the gaseous state.
  B)   As the pressure in a glass cylinder containing pure CO_2 is increased from 65 to 80 atm, no meniscus or interface between the liquid and gas phases is observed.
  C)   The solid, liquid, and gas states co-exist.

4.32 Sulfur has two crystalline forms, rhombic and monoclinic sulfur which represent different arrangements of the S_8 rings (see Problem 4.13). A phase diagram for S is shown on the right.

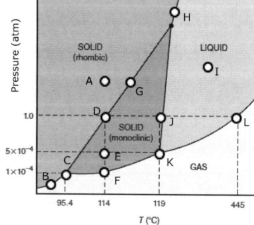

  A)   For each lettered point, give the:
     (1)   Number of phases present
     (2)   Degrees of freedom
     (3)   If an equilibrium, what phases are in equilibrium
     (4)   If a single phase, what phase is present
  B)   What actions would you need to take to go from $D \rightarrow G$?
  C)   Describe what you would observe in the sulfur sample when you go from the points $D \rightarrow J \rightarrow L$.
  D)   What is the relationship between μ_s, monoclinic and μ_s, rhombic or μ_{liquid} as you go from $A \rightarrow D \rightarrow E \rightarrow F$ points?
  E)   What is the significance of Point C at $T = 95.4°C$ versus Point K at $119°C$?
  F)   What is the significance of $T = 445°C$?
  G)   What point, if any, on the diagram would be considered the normal melting point of S(s)?

4.33   Consider the phase diagram for CO_2 given below, and that you have a sample of CO_2 in a glass tube that allows you to see the state(s) of CO_2 at the different T and P values. If you start at the blue circle and make the changes indicated on the diagram, state:

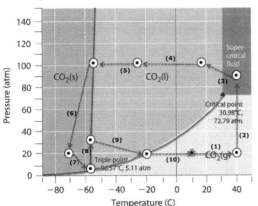

A) For the shifts (1)–(10), identify which are:
    (a) Isothermal changes?
    (b) Isobaric changes?
    (c) Neither isobaric or isothermal?
B) Which changes would result in a change in what you observe in the tube and what would the change be.
C) When would a boundary between phases be visible?

4.34   Using some of the data you determined for naphthalene, $C_{10}H_8$, in Problem 4.27 and the additional information given below, sketch the phase diagram for naphthalene in the box provided, identifying the three phase regions with the appropriate equilibrium lines, the normal melting point, normal boiling point, triple point and start of the "supercritical region".

Density solid $= 1.025$ g/cm³
Normal $T_{bp} = 217°C$
Density liquid $= 0.9625$ g/cm³
Normal $T_{fp} = 80.2°C$
$P_c = 41.4$ atm
$T_c = 475°C$

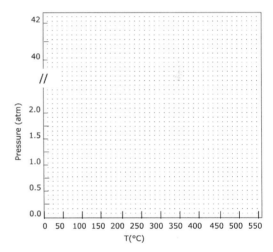

4.35   State whether the following statements are TRUE or FALSE. When false, identify the error(s) in the statement:
A)   At equilibrium, a substance that occurs in two phases will have same number of moles in both phases.
B)   The entropy for a phase transition can be positive or negative, but never zero.
C)   The Clausius–Clapeyron equation can be used for a solid–liquid phase transition ONLY.

4.36   In terms of the change in chemical potential in responses to temperature and pressure changes for a phase diagram,
A)   Why are liquid ⇔ solid equilibrium lines linear? What determines whether or not they have either positive or negative slopes?
B)   Why do liquid ⇔ vapor lines always have a positive slope and always show a curved line?
C)   What has to be true about the chemical potential for any two phases that are at equilibrium?
D)   What is the:
    (a)   Significance of the point where the three equilibrium lines cross?
    (b)   What is true about the chemical potential of the three states at that point?
    (c)   Why are there no degrees of freedom at this point?

KEY POINTS – THE CHEMICAL POTENTIAL OF REAL GASES AND FUGACITY

Fugacity, f, is a measure of the vapor pressure when the vapor acts as a real gas, rather than as an ideal gas, and is the effective pressure exerted.

$$f = \text{effective pressure} = \phi P$$

For gases $\quad \overline{G}(P_2) = \overline{G}(P_1) + RT \int_{P_1}^{P_2} \frac{dP}{P}$

If not acting as ideal gas can adjust using fugacity, f, and the compressibility factor, Z:

f, fugacity $= \phi \times P$

$\ln\phi = \int_0^P \frac{Z-1}{P} dP$

Van der Waals $\quad \ln\phi = \frac{bP}{RT} - \frac{aP}{(RT)^2}$

Virial Equation $\quad \ln\phi = \frac{BP}{RT}$

ϕ is defined as the "fugacity coefficient" and depends on chemical identity, T and P (like Z).

- It depends on the same factors as Z
- It can be greater than 1.0 or less than 1.0, like Z, depending on which type of interactions, attractive or repulsive, are dominating
- The value of ϕ will vary with pressure and temperature
- It has no units (is "dimensionless")
- Will become equal to 1.0 as $P \to 0$

The chemical potential for a real gas then becomes:

$$\mu_i = \mu_i^\circ + RT \ln \frac{f}{P^\circ}$$

So that:

$$\mu_i = \mu_i^\circ + RT \ln \frac{f}{P^\circ} = \mu_i^\circ + RT \ln \frac{P}{P^\circ} + RT \ln \phi = \mu_i^\theta + RT \ln \phi$$

where $P^\circ = 1$ atm (or 1.0 bar) and μ_i^θ is the ideal gas value.

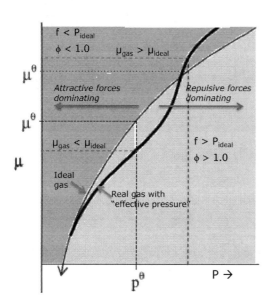

- *The adjustment for a real gas is $RT \ln\phi$ and its effect on the chemical potential is shown in the figure on the right*
- The relationship between Z, the compressibility factor, and ϕ is defined as: $\ln\phi = \int_0^P \left(\frac{Z-1}{P} \right) dP$

Since both the van der Waals and virial equations have definitions for Z, these can be substituted for Z in the integral and the integration, then produce simple equations for the determination of $\ln\phi$.

Van der Waals	Integral Becomes:	Integrated Result:
$Z_{VdW} = 1 + \left(b - \frac{a}{RT} \right) \frac{P}{RT}$	$\ln\phi = \int_0^P \left(\left(b - \frac{a}{RT} \right) \frac{1}{RT} \right) dP$	$\ln\phi = \frac{bP}{RT} - \frac{aP}{(RT)^2}$
Virial Equation	Integral Becomes:	Integrated Result:
$Z_{\text{Virial}} = 1 + \frac{B}{V_m} = 1 + B\left(\frac{P}{RT} \right)$	$\ln\phi = \int_0^P \left(\frac{B}{RT} \right) dP$	$\ln\phi = \frac{BP}{RT}$

Further approximations can be made under certain conditions:

(1) If repulsive interactions are dominant, the first factor in the van der Waals equation is the most important factor and a shortened equation can be used: $\ln\phi = \frac{bP}{RT}$

(2) If Z is close to 1.0, the approximation $f = Pe^{(Z-1)} \approx PZ$ applies where $Z = \phi$.

EXAMPLE PROBLEMS

4.37 From the definition of ln ϕ we know that Z and the fugacity coefficient are definitely related to one another. The graphs below show the behavior of Z with changes in pressure and also how the fugacity coefficient varies with pressure for the same gases.

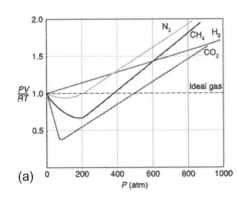

(a)

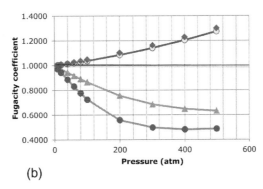

(b)

A) Compare the behavior of the gases N_2, H_2, and CO_2 on the two graphs.
   (a) The Z value for H_2 is generally greater than 1.0 as P increases; what is true about its fugacity coefficient?
   (b) The behavior of H_2 and N_2 are very similar on the fugacity coefficient graph, but very different from that of CO_2. Is this consistent with the Z values for the gases? Explain.

B) Why is the trend for the fugacity coefficient ϕ for CO_2 and NH_3 so different from that of H_2 (or N_2)? What is different about the interactions between the gas particles that can explain this behavior?

C) Given that $\mu_{real\ gas} - \mu_{ideal}^\theta = RT\ln\phi$, what will be true about $\Delta\mu$ when ϕ is greater than 1.0? What about when ϕ is less than 1.0?

4.38 Given that for $CO_2(g)$ at 150°C and 20 atm, the ln ϕ is –0.03379, and at 150°C and 200 atm, it has a value –0.3378, calculate the:
   A) Fugacity, f, of CO_2 at 20 atm and 150°C. What is the % difference from the ideal P value?
   B) Fugacity, f, of CO_2 at 200 atm and 150°C. What is the % difference from the ideal P value?
   C) Determine the $\Delta\mu$ value of the gas, defined as $\mu_{real} - \mu_{ideal}$ at 20 atm, 150°C and then 200 atm, 150°C. Is the μ_{real} gas value increasing, decreasing, or staying the same?

4.39 Given the equations for ln ϕ for both the van der Waals and virial equations in the table above:
   A) Prove that the units on the righthand side of either ln ϕ equation will cancel, so that ln ϕ is dimensionless.
   B) Determine the values of ln ϕ, ϕ and the fugacity (effective pressure) when $NH_3(g)$ is at 25°C, 150°, and 300°C while $P = 10$ atm, using the van der Waals and virial equations for ln ϕ. The values of the second virial coefficient for the 3 temperatures are –265, –100, and –45.6 cm³/mol, respectively. [A spreadsheet calculation would be useful for this analysis.]
   C) Then answer the following questions:
      (a) How well do the results from the two equations agree? (Be specific!)
      (b) Is $\Delta\mu$ increasing, decreasing or staying the same as T increases at 10 atm? What could be the main factor(s) contributing to this trend?
      (c) When you consider the effective pressure, is $NH_3(g)$ becoming more like an ideal gas as the T increases, or less?

4.40 To examine the effect of T on the fugacity coefficient, consider N_2 gas that has the values for B, the second virial coefficient, given below at five different temperatures.

T (K)	200	273	350	450	550
B (cm³/mol)	−35.5	−10.8	2.7	12.7	18.8

A) Before doing any calculation, state what you think will be the effect on the fugacity coefficient, ϕ, of B going from negative values to positive values at the different temperatures?

B) Create a table (using a spreadsheet) of $\ln \phi$, $\Delta\mu$ and ϕ for $N_2(g)$ at the five temperatures when P is held constant at 100 atm, using the estimation of $\ln \phi$ from the virial equation.

C) To consider the effect of lowering P, construct two more tables of $\ln \phi$, $\Delta\mu$ and ϕ for $N_2(g)$ at the five temperatures but for a pressure of 20 atm and then 2 atm. Compare the results to the values in (B).

    (a) Are the values of $\Delta\mu$ getting larger or smaller as the P is lowered?

    (b) Given the results of $\Delta\mu$, at what temperatures is $\mu_{real} > \mu_{ideal}$, $\mu_{real} < \mu_{ideal}$, or $\mu_{real} = \mu_{ideal}$ for this pressure?

    (c) Which forces are dominating at each temperature when $N_2(g)$ or are they balanced as in an ideal gas?

    (d) The figure in the Key Points section above shows the possible variation of μ with P, but, based on your results for the overall behavior of μ for N_2, could the same type of figure be drawn for μ_{real} and μ_{ideal} versus T?

Free Energy (ΔG) of Mixing, Binary Liquid Mixtures, Colligative Properties, and Activity

5

KEY POINTS – FREE ENERGY OF MIXING

Mixing of two or more gases (A, B, …) is always spontaneous since, for gases, ΔG_{mix} is always negative. if the gases are ideal, then $\Delta H_{mix} = 0$ and $\Delta G_{mix} = -T\Delta S_{mix}$ and can be defined as shown on the map. For two-component systems:

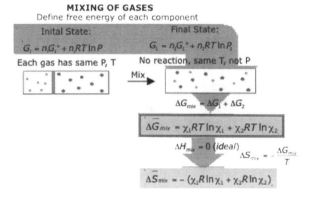

MIXING OF GASES
Define free energy of each component

Inital State:
$G_i = nG_i° + n_iRT \ln P$
Each gas has same P, T

Final State:
$G_L = n_LG_L° + n_LRT \ln P_L$
No reaction, same T, not P

Mix

$\Delta G_{mix} = \Delta G_1 + \Delta G_2$

$\Delta \bar{G}_{mix} = \chi_1RT \ln \chi_1 + \chi_2RT \ln \chi_2$

$\Delta H_{mix} = 0$ (ideal) $\Delta S_{mix} = -\dfrac{\Delta G_{mix}}{T}$

$\Delta \bar{S}_{mix} = -(\chi_1R \ln \chi_1 + \chi_2R \ln \chi_2)$

- ΔG_{mix} varies with the mole fraction of the component gases, χ_A and χ_B.
- Since ΔS_{mix} is always positive, and T in Kelvin is always positive as well, ΔG_{mix} will always be a negative value.
- The mole fraction of each gas sets the value of ΔS_{mix}, but, for ΔG_{mix}, both T and the mole fraction will determine the value. (Consult the map.)
- The maximum value of ΔS_{mix} and the minimum value of ΔG_{mix} occurs when the mole fractions of the gases are the same.

If the gases are acting as real gases, then $\Delta H \neq 0$, but ΔH_{mix} should be small in comparison to $T\Delta S_{mix}$, so that ΔG_{mix} is still determined by $-T\Delta S_{mix}$.

EXAMPLE PROBLEMS

5.1   Laughing gas, which is commonly used in dental offices as a quick-acting anesthetic, is a 50/50 mixture (v/v) of $N_2O(g)$ (commonly called nitrous oxide) and $O_2(g)$, that is sold in tanks of various sizes. No reaction takes place between the two gases, so it is very stable. If the $T = 22°C$ and the pressure inside the tank is 10.0 atm, what would be the
   A)   (a) ΔG_{mix}?   (b) ΔS_{mix}?
          Furthermore
   B)   If the pressure in the tank decreases to 2.0 atm, would this affect the values in A)?

5.2   A mixture used to treat hyperventilation is a combination of 20.0 L O_2, 20.0 L He and 4.0 L CO_2 gases, with each gas having a $P = 1.0$ atm and T of 25°C.
   A)   What is the mole fraction of each gas in the mixture?
   B)   Calculate n_{total} and ΔG_{mix} in terms of kJ.

5.3   Nitrox I is a breathing mixture for deep sea diving that is 32.0% O_2 and 68% N_2 (v/v). Normal air, that contains 21.0% $O_2(g)$ and 79% $N_2(g)$ (v/v), would contain too much N_2 and would cause nitrogen narcosis at the higher pressure that divers experience.

A)   How many liters of pure $O_2(g)$ at 20 atm and 25°C should be added to 100 L of air, also at 20 atm and 25°C, to change the % O_2 from 21% to 32%?

B)   (a)   Would you expect the ΔG_{mix} to increase or decrease after the O_2 is added? Briefly, give the reason(s) for your choice.

   (b)   Determine the difference between ΔG_{mix} for Nitrox I and ΔG_{mix} for air and determine whether or not your expectation stated in (a) was met.

5.4   Trimix is gaseous mixture used for scuba diving to extreme depths, that contains 26% O_2, 17% He and 57% N_2 (w/v). The partial pressure of O_2 in the mixture must be between 100 and 120 kPa to prevent oxygen toxicity. Suppose the partial pressure of O_2 is 120 kPa, what is:

A)   The total pressure of the mixture in kPa and atm?

B)   The partial pressure of He and N_2 in kPa and atm?

C)   The ΔG_{mix} of Trimix at 25°C?

5.5   Consider a 5.0-L container that is divided into two chambers, where one contains 3.0 L of $N_2(g)$ separated by a partition from 2.0 L of $Xe(g)$ in the second chamber, with both gases being at 1.0 atm and 25°C.

A)   If the partition were removed, what would be the ΔG_{mix} and ΔS_{mix}?

B)   Instead of having equal pressures, suppose the pressure of the 3.0 L of N_2 gas was 4.0 atm and that for Xe was 2.0 atm, but both gases at were still at 25°C. If the partition were removed,

   (a)   Would the ΔG_{mix} and ΔS_{mix} be the same as in A)? If not, what would the values be?

   (b)   What other factors would have to be included to describe ΔG for the process?

C)   In terms of ΔG_{total}, prove that the mixing and pressure changes will still be a spontaneous process.

KEY POINTS – ΔG AND TWO-COMPONENT LIQUID MIXTURES

The solutions considered here are homogeneous mixtures, having a constant composition throughout, and with at least two different components, which are in equilibrium with each other.

- In a binary mixture, the component with the greatest number of moles is the solvent, A. The solvent sets the physical properties of the solutions, such as boiling point, volume, density and melting point, as long as it is the major component. The minor component, the solute, is represented as B.
- Unlike gas mixtures, when two liquids are mixed, there must be interactions between the A molecules, A···A, and the B molecules, B···B, that occur in the pure liquids.

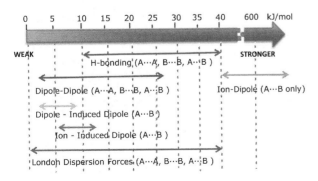

For a solution to form, A···B interactions need to be approximately equal to or greater than A···A. Non-covalent interactions (or intermolecular forces) are attractions between molecules or ions and are weaker than chemical bonds. They exist in pure substances as well as solutions. The major types and their relative strength are shown in the figure on the right. The types of interactions that are unique to solution and not found in pure substances would be ion-dipole interactions.

In an "ideal solution", the A⋯A forces are very similar to the B⋯B forces, so that the environment around each molecule is similar to its pure state. In that case, ΔG_{mix} and ΔS_{mix} would be defined in the same way as for gases.

$$\Delta G_{mix}, \text{ideal} = RT\left[\chi_A \ln \chi_A + \chi_B \ln \chi_B\right] \quad \Delta H_{mix}, \text{ideal} = 0$$

$$\Delta S_{mix}, \text{ideal} = -R\left[\chi_A \ln \chi_A + \chi_B \ln \chi_B\right]$$

In real solutions, ΔH_{mix} often has a finite value, depending on the magnitude of intermolecular forces or non-covalent interactions within each component, A⋯A and B⋯B. These interactions have different strengths (as described above) and can induce changes in the organizational structure of the liquid molecules, which in turn affects both ΔH_{mix} and ΔS_{mix}.

■ The minor component by moles, B, called the solute, will interact mainly with A. If the A⋯B interactions are weaker than A⋯A, then the two liquids will not mix and are immiscible. There would be two distinct phases in the regions, where B is immiscible in A.
■ In order for A and B to form a mixture, the A⋯B interactions would need to be greater in strength than either A⋯A or B⋯B, so that a solution forms.
■ The major thermodynamics values for the solution, ΔG_{mix}, ΔS_{mix}, and ΔH_{mix}, will vary with the mole fractions of A and B and can only be defined at a specific composition of the solution.

KEY POINTS – REAL SOLUTIONS AND EXCESS THERMODYNAMIC FUNCTIONS FOR MIXTURES

In an ideal solution where A⋯B ≈ A⋯A ≈ B⋯B, the volumes add as expected and:

$$V_{mix} = n_A \overline{V}_A + n_B \overline{V}_B \text{ so that } \overline{V}_{mix}(\text{ideal}) = \chi_A \overline{V}_A + \chi_B \overline{V}_B$$

In the mixture, A and B each has its own **partial molar volume**,

$$\overline{V}_A = \left[\frac{\partial V}{\partial n_A}\right]_{P,T,n_B} \quad \overline{V}_B = \left[\frac{\partial V}{\partial n_B}\right]_{P,T,n_A}$$

But when the interactions are not the same strength, then the molar volume of each component will vary, as, at first, B disrupts A⋯A interactions and A interferes with B⋯B interactions. The actual volume of solution can be either greater than or less than the ideal solution value. As the forces around the A and B molecules change, the partial molar volumes can either increase or decrease.

Differences between the real volume of the solution and the expected value from the ideal solution indicates that other major thermodynamic properties of the mixture will also be affected. The difference between the thermodynamic values of the real or "regular" solution and the ideal values can be measured and are denoted by using the thermodynamic symbol with a superscript "E", for "excess". An "excess" thermodynamic property is defined as the difference between the thermodynamic property of mixing for the system concerned and that for an ideal one at the same temperature, pressure, and composition.

There are four "excess" thermodynamic properties for real solutions – volume, enthalpy, entropy and free energy of mixing, as defined below. The importance of the parameters is that they provide knowledge of the nature of the molecular interactions A⋯B and how the interaction changes with composition, on the nanoscale in the binary mixture, that are very difficult to measure directly or to predict.

Excess Volume of Solution	Excess volume is the difference between the actual volume of the solution and the expected "ideal" volume if the molar volumes stayed constant at the values for the pure liquids.
$V^E = \Delta \bar{V}_{mix} = \dfrac{V_{soln}}{n_A + n_B} - (\bar{V}_A^{\,*} + \bar{V}_B^{\,*})$ $\bar{V}_{soln} = \bar{V}_{ideal} + \Delta \bar{V}_{mix} = \bar{V}_{ideal} + V^E$	
Excess Enthalpy $H^E = \Delta \bar{H}_{mix}$	Since the enthalpy of mixing an ideal solution is zero, any enthalpy change measured when a solution is formed is considered to be the excess enthalpy of mixing. <ul><li>A positive (endothermic) H^E indicates that the A⋯B interactions are weaker than A⋯A or B⋯B, depending on which is the solvent.</li><li>A negative (exothermic) H^E indicates that the A⋯B interactions are stronger than A⋯A or B⋯B.</li></ul>
Excess Entropy $S^E = \Delta \bar{S}_{mix} - \Delta \bar{S}_{ideal}$ $= \Delta \bar{S}_{mix} - \left[-R(\chi_A \ln \chi_A + \chi_B \ln \chi_B) \right]$ $= \Delta \bar{S}_{mix} + R\left[\chi_A \ln \chi_A + \chi_B \ln \chi_B \right]$	Excess entropy changes arise from the way the A and B molecules may cluster together in the mixture or change the overall ordering, instead of moving freely, or maintain the arrangement of the ideal solution. <ul><li>A positive S^E indicates that means disorder is increasing as B is added to A ($\chi_B < 0.5$) or A added to B ($\chi_B > 0.5$), so that the normal organization of either is being disrupted.</li><li>A negative S^E indicates an increase in ordering, due to complex formation or some type of clustering of A and B molecules, that does not exist in their pure liquid states.</li></ul>
Excess Free Energy $G^E = \Delta \bar{G}_{mix} - \Delta \bar{G}_{ideal}$ $= \Delta \bar{G}_{mix} - RT\left[\chi_A \ln \chi_A + \chi_B \ln \chi_B \right]$ $\Delta \mu_{mix} = \Delta \bar{G}_{mix} = \mu_{(soln)} - \mu_{mix}^{*}$ $= \left[\mu_{A(soln)} + \mu_{B(soln)} \right] - \left[\mu_A^{*} + \mu_B^{*} \right]$	The excess enthalpy and entropy values will then produce an "excess" free energy for the solution formation that affects the spontaneity of mixing. $\Delta \bar{H}_{mix} = \Delta \bar{G}_{mix} + T \Delta \bar{S}_{mix} \Rightarrow H^E = G^E + TS^E$ or $G^E = H^E - TS^E$ <ul><li>G^E can be either positive or negative, where negative values would enhance the mixing of the components versus the ideal.</li><li>If the H^E is endothermic, and S^E is negative, then G^E is positive and may be large enough so that A and B will not mix. This explains the observation that, if the A⋯B interactions are weak compared to A⋯A, the two liquids are immiscible.</li></ul>

The following example problems will show how knowledge of the excess properties can lead to a better understanding of the interactions between A and B in the mixtures. They are often very complex and difficult to predict, so that only by measuring the actual properties can theories about the molecular interactions or arrangements be developed.

EXAMPLE PROBLEMS

5.6 Liquid hexane, C_6H_{14}, ($d = 655$ kg/m³) and liquid heptane, C_7H_{16}, ($d = 684$ kg/m³) form an ideal mixture. In order to achieve the most spontaneous mixing, what proportions of hexane and heptane should be mixed in terms of:
   A) Moles of each liquid?
   B) % mass of each liquid?
   C) Volumes of each liquid needed to make a total volume of 250 mL, assuming that no contraction or expansion occurs since the mixture is an ideal mixture?

5.7 The two graphs below illustrate the change in molar volume, the "excess volume", ΔV_{mix}, for a mixture of methanol (CH_3OH), as B, and water, as A, at 20°C. The first graph shows how the actual ΔV_{mix} varies with mole fraction of B (CH_3OH), and the second graph shows how the individual partial molar volumes of A (H_2O) and B (CH_3OH) vary with χ_B.

A)   Note how the solution volume is smaller than expected if the solution were ideal, as the χ_B goes from zero to 0.5. Using intermolecular forces, explain what is happening at the molecular level to cause the contraction of the solution.

B)   The method of intercepts that can be used to determine the separate molar volumes of A (H_2O) and B (CH_3OH) is illustrated as well. The black dashed line, the tangent to the line at the open black circle, gives the separate values of $\overline{V}_A$ and $\overline{V}_B$ for the mole fraction (≈ 0.30) CH_3OH.

    (a)   Given the density of water and methanol are 0.9982 and 0.7872 g/cm³, respectively, calculate the molar volumes for pure water, $\overset{\bullet}{\overline{V}}_A$, and methanol, $\overset{\bullet}{\overline{V}}_B$.

    (b)   Prove that the values of V_i marked by black open circles on the second graph represent the partial molar volumes of A (H_2O) and B (CH_3OH) in the solution.

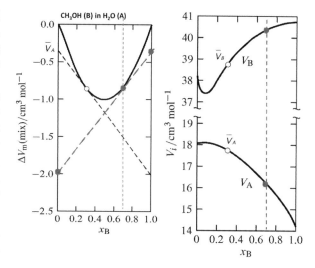

C)   Using the same approach as in (B), and the tangent to the curve at the red circle, $\chi_B = 0.70$, represented as the red dashed line, determine the:

    (a)   Change in molar volume of the solution, ΔV_{soln}, at the red (or gray) circle, with the tangent represented by the dashed line.

    (b)   Two individual intercepts, ΔV_A and ΔV_B, from the intercepts to the tangent line at $\chi_B = 0.70$ from the first graph.
    Furthermore

    (c)   Prove the partial molar volumes of A (H_2O) and B (CH_3OH), marked by red (or gray) circles on the second graph, $\chi_B = 0.70$, correspond to the intercept values being applied to the pure molar volumes of A and B.

5.8   The excess molar volume, ΔV_{mix}, was measured for mixtures of isopropanol (2-propanol, $CH_3CHOHCH_3$) and water at 25°C, as shown on the right.

A)   (a) Is the solution contracting or expanding as the isopropanol is mixed with water?

    (b)   Compare the types of forces A···A, B···B to A···B. Why would a solution be expected to form? What does this indicate about the sign of the ΔG_{mix} overall for the solution formation?

    (c)   What does the observed change in molar volume tell you about what is happening on the molecular scale in the solution?

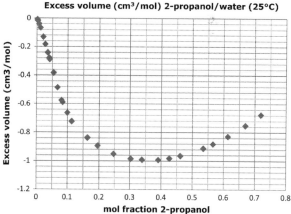

B)   When $\chi_B = 0.3029$, the excess volume was measured as -0.9842 cm³/mol. Given the density of pure isopropanol is 0.7804 g/cm³ and the partial molar volume for pure water is 18.054 g/cm³:

    (a)   What is the ideal molar volume of the solution?

    (b)   What is the actual molar volume of the solution?

    (c)   If the partial molar volume of water is 17.154 cm³/mol, what is the partial molar volume of isopropanol at $\chi_B = 0.3029$? How does this compare to its pure molar volume?

5.9   When ethanol and water are mixed, the solution volume is always less than the ideal volume, as the figures shown below illustrate. The individual partial molar volumes of water and ethanol change significantly, as the second graph shows, as the ratio of ethanol to water molecules changes.

A)   Calculate the ideal molar volume of pure water and ethanol, given that the densities are 0.9982 g/cm³ and 0.7893 g/cm³, respectively.

B)   (a)   For what range of $\chi_{ethanol}$ does the partial molar volume of water exceed its pure value?

       (b)   For what range of $\chi_{ethanol}$ does the partial molar volume of ethanol exceed its pure value?

C)   (a)   What would the expected A···B interactions be in the mixture?

       (b)   Would the expected A···B interactions change with composition in the mixture?

D)   Considering that liquid water has an extensive 3-D H-bonded network that keeps larger spaces between the water molecules than the atoms require, what might be an explanation of the anomalies seen in the partial molar volumes at $\chi_{ethanol} \approx 0.10$? If only the excess volume of solution plot were shown, would you be aware there was an anomaly at $\chi_{ethanol} \approx 0.10$?

(a) V^E ethanol/water

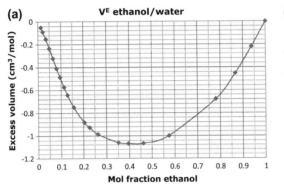

(b)

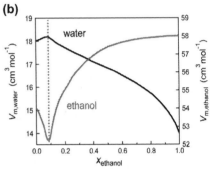

5.10   From the data given in Problem 5.8, the maximum contraction of the solution occurs at $\chi_{ethanol} = 0.400$. The partial molar volumes of water and ethanol, C_2H_5OH, at $\chi_{ethanol} = 0.400$ were measured as 17.4 and 56.5 cm³/mol, respectively. Suppose you wanted to prepare this solution starting with 50.00 mL of pure water:

A)   What volume of pure ethanol ($d = 0.7893$ g/cm³) should be added to the water to produce a solution with $\chi_{ethanol} = 0.400$?

B)   Calculate the total contraction or "excess" ΔV_{mix} in mL (cm³) of the solution.

C)   Is your answer to A) consistent with the value shown in the graph of excess volume for a solution with $\chi_{ethanol} = 0.400$?

D)   Would you be able to notice the contraction if you were making the mixture in a 200-mL graduated cylinder with 5.00-mL markings?

5.11   The density of various liquid mixtures of isopropanol, $(CH_3)_2CHOH$ [MW 60.096], and methanol, CH_3OH [MW 32.042], have been measured at 25°C [Dortmund Data Bank Set 5043]. If a mixture of 64.084 g CH_3OH and 43.65 g $(CH_3)_2CHOH$ is found to have a density of 0.7851 g/cm³:

A)   What is the actual volume of the solution in cm³?

B)   Given that the densities of pure CH_3OH and $(CH_3)_2CHOH$ are 0.7872 g/cm³ and 0.7804 g/cm³, respectively, what is the:

       (a)   Partial molar volume of each pure component?

       (b)   Total volume of the solution, if ideal, and ΔV_{mix} in cm³? Furthermore,

       (c)   If you were measuring the solution volume in a 150-mL graduated cylinder as you mixed it, would you be able to see the difference in expected (ideal volume) versus the actual volume, if the cylinder was graduated in 1.00-mL increments?

C)   (a)   How does the actual density of the solution compare to the density of an ideal solution?

       (b)   Does this indicate that the solution has contracted or expanded versus an ideal solution?

       (c)   Considering the data given and the intermolecular forces and structures of the molecules, explain why an ideal solution has or has not formed between the two components.

5.12 The graphs on the right show the data for the excess molar volume and excess enthalpy for the mixture of acetone (CH_3COCH_3) and water (H_2O) [Dortmund Data Bank Sets 4132, 1407].

A) What are the dominant intermolecular forces in:
   (a) Pure water as A···A?
   (b) Pure acetone as B···B?
   (c) The mixture as A···B?

B) Concerning the excess volume (first) graph:
   (a) What does the graph indicate about the solution volume versus the ideal volume?
   (b) At what molecular ratio does the maximum effect occur?

C) In the second graph, there are two distinct regions on the excess enthalpy, H^E (or ΔH_{mix}) graph.
   (a) Which component of the solution is the solvent when H^E is exothermic?
   (b) At what molecular ratio does the maximum effect occur?
   (c) What do exothermic values indicate about the strength of A···B interactions versus A···A? Is this consistent with some type of cluster forming in the solution between acetone and water?

D) (a) Which component of the solution is the solvent when H^E is endothermic?
   (b) What do endothermic values indicate about the strength of A···B interactions versus A···A? Is this result consistent with the cluster forming only when mol fraction of acetone is low?

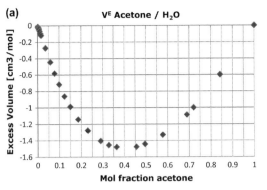

(a) V^E Acetone / H₂O

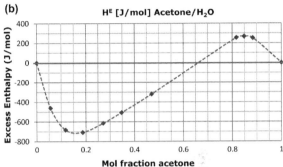

(b) H^E [J/mol] Acetone/H₂O

5.13 The graphs on the right show the data for the excess molar volume and excess enthalpy for the mixture of acetone (CH_3COCH_3) and chloroform ($CHCl_3$) [Dortmund Data Bank Sets 2002, 417]. Note how the ΔV changes sign as the mole fraction of acetone increases, which is different behavior than that seen for the binary solutions in the previous problems. The second graph shows the excess enthalpy, H^E [or ΔH_{mix}], observed for the mixture.

A) What are the dominant intermolecular forces in:
   (a) Pure chloroform as A?
   (b) Pure acetone as B?

B) This time, there are two distinct regions on the excess volume graph:
   (a) What does the graph indicate about the solution volume versus the ideal volume?
   (b) At what molecular ratio does the maximum contraction occur?

C) The excess enthalpy, H^E (or ΔH_{mix}) graph does not show two regions but is exothermic at all mole fractions.
   (a) At what molecular ratio does the maximum effect occur?
   (b) What do exothermic values indicate about the strength of A···B interactions versus A···A or B···B?

D) Measurement of certain properties of the binary mixture indicates that CH_3COCH_3 and $CHCl_3$ are forming a 1:1 complex via H-bonding (as shown in the figure on the right) between the O in C=O and the H on C in $CHCl_3$.
   (a) How do the H^E data support that theory?
   (b) Why would this be a very unusual H-bond?

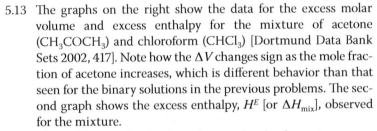

(a) V^E(mL) CHCl₃/acetone

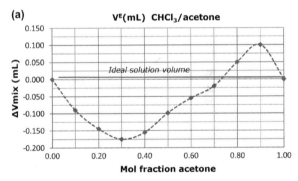

(b) H^E [J/mol] CHCl₃/acetone

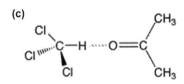

(c)

5.14 The graphs on the right show H^E, G^E, and S^E for six binary mixtures, with the components indicated. If the mixtures were ideal, the values of ΔG_{mix} and $T\Delta S_{mix}$ at 25°C would be the same, but none are. Each mixture also shows a very different pattern for H^E and S^E, indicating that there are different kinds of interactions occurring between the molecules in the mixture.

[Adapted from: H.C. Van Ness and M.M. Abbott, Perry's Chemical Engineering Handbook (7th edition) McGraw-Hill 1997]

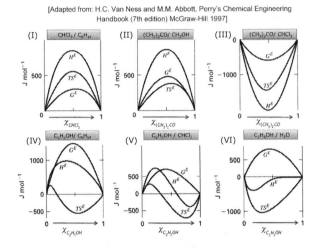

A) What would be the maximum values for ΔG_{mix} and $T\Delta S_{mix}$ in joule/mol, if the mixtures were ideal mixtures?

B) Considering the definition given for G^E in the Key Points earlier:
   (a) Estimate the value of ΔG_{mix} observed for mixtures I, III, and IV.
   (b) Should the two substances spontaneously mix in each mixture?

C) Compare the mixtures indicated in terms of:

(1) The observed sign and magnitude of H^E	(2) What this indicates about the A⋯B interactions versus that in pure solvent in each mixture	(3) The trend(s) for S^E, and what it indicates about the change in disorder as the mole fraction of solute increases in the mixture.

   (a) Mixtures I and IV (n-hexane mixed with chloroform or ethanol)
   (b) Mixtures II and III (acetone mixed with chloroform and methanol)
   (c) Mixtures V and VI (ethanol mixed with chloroform and water)

5.15 The excess enthalpy shows a T dependence, as would be expected of thermodynamic properties. The figures below show the T dependence of the acetone (CH_3COCH_3) and water binary system, as well as that of acetonitrile (CH_3CN) and water.

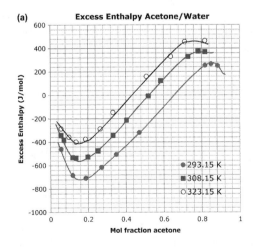

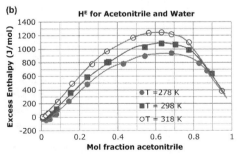

A)   Although the basic pattern is maintained, what changes do you see in the H^E curves for each system?

B)   What equation, introduced in Chapter 2 and on your map, would define the T dependence of H^E?

C)   Based on that equation, what term could we determine from the T change of the H^E values?

5.16   The excess functions can be used to define many physical responses of the binary systems. The excess volume defines the pressure dependence of G^E, while H^E defines the T dependence of μ_{mix} or G^E. Two equations that express the relationships were given as:

$$\frac{V^E}{RT} = \left[\frac{d(G^E/RT)}{dP}\right]_T \quad \text{and} \quad \left[\frac{d(G^E/RT)}{dT}\right]_P = -\frac{H^E}{RT^2}$$

A)   What definitions (equations introduced in Part 3) are the basis for these two equations?

B)   What is the benefit in dividing G^E and H^E by R in the second equation?

C)   What will need to be included when applying the equation to make sure that the units on both sides of the equal sign are the same?

D)   Apply the equations to determine the P and T dependence of the acetone and chloroform system, when $\chi_{(CH_3)_2CO} = 0.40$ where $V^E = -0.15$ cm³/mol and $H^E = -1900$ J/mol (Problem 5.12). How does the pressure dependence compare to the temperature dependence?

KEY POINTS – VAPOR PRESSURE, RAOULT'S LAW, AND HENRY'S LAW FOR MIXTURES

One of the first obvious physical properties affected by creating mixtures is the vapor pressure over the solution. Vapor pressure, or conversion of liquid molecules to the gas phase, occurs only at the surface of the liquid below the boiling point. At the boiling point, when the vapor pressure equals atmospheric pressure, molecules within the bulk of the liquid convert to gas, creating the gas bubbles characteristic of boiling. At the boiling point, the vapor pressure of the liquid equals the atmospheric pressure and molecules within the bulk of the liquid convert to gas, creating the bubbles characteristic of boiling.

The fact that there are fewer A and B molecules at the surface of the solution relative to the pure liquids, leads to a lowering of the vapor pressure of each liquid in proportion to its mole fraction at the surface, as defined by **Raoult's law**, $P_i = \chi_i P_i^\circ = \chi_i P_i^*$ where P_i° (or P_i^*) equals the pure vapor pressure value for the component. The total pressure over a binary solution is then the sum of the vapor pressures of A and B.

$$P_{\text{total}} = P_A + P_B = \chi_A P_A^* + \chi_B P_B^* \text{ or since } \chi_A + \chi_B = 1.0, \ P_{\text{total}} = (1-\chi_B)P_A^* + \chi_B P_B^* = P_A^* + \chi_B\left[P_B^* - P_A^*\right]$$

An ideal solution, where the A⋯B forces are very similar to those of A⋯A (or B⋯B), obeys Raoult's law and produces a plot of pressure versus mole fraction like that shown below, with straight lines giving the separate values of P_A, P_B, and P_{total} over the solution. The graphs are called **binary liquid–vapor phase diagrams**.

In real (or regular solutions), where the A⋯B forces are not similar to those of A⋯A (or B⋯B), such as those discussed in the earlier section, will not show straight lines, but instead curves, as either positive or negative deviations from Raoult's law where the value changes with composition (as shown below).

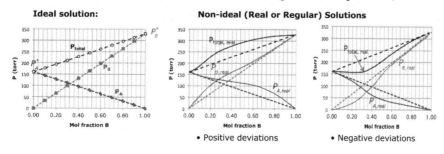

• Positive deviations   • Negative deviations

Positive deviations from Raoult's law occur when:

■ A···B forces are weaker than A···A or B···B forces.
■ Less energy is required for molecules to leave the solution than the pure liquid.
■ The excess enthalpy, H^E, is endothermic for the solution indicating A···B < A···A.

Negative deviations from Raoult's law occur when:

■ A···B forces are stronger than A···A or B···B forces.
■ More energy is required for molecules to leave the solution than in the pure liquid.
■ The excess enthalpy, H^E, is exothermic for the solution indicating A···B > A···A.

KEY POINTS – HENRY'S LAW FOR BINARY MIXTURES

In a non-ideal mixture, at low mole fraction values of the solute, B:

■ The solvent A generally acts as if it is in an ideal solution and obeys Raoult's law, since it mainly interacts with itself.
■ If the deviations are not too large at low mole fractions of B, P_B also appears linear and predicts an intersection on the y-axis that is higher or lower than P_B^*.
■ This is called the **Henry's law region for B** (see figure on the right). The point where the line intersects the y-axis. when χ_B is 1.0, is the Henry's law constant, $K_{H,B}$, for the solute B in the solvent A.
■ The Henry's law constant, $K_{H,B}$ then replaces P_B^* in Raoult's law and

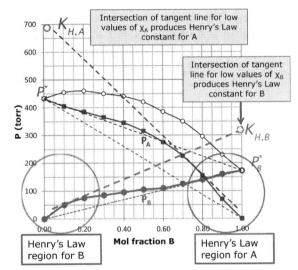

produces Henry's law: $P_B = \chi_B K_{H,B}$ where $K_{H,B}$ has pressure units and predicts the value of P_B at low mole fractions of B.

At high mole fractions of solute, B:

■ The solute B obeys Raoult's law, since the B···A interactions are outnumbered by B···B interactions.
■ If the deviations are not too large, the P_A values also follow a linear relationship, but with an intersection on the P axis that is not equal to P_A^*, which is called the Henry's law region for A.
■ The intersection on the y-axis of the line that describes P_A at high mole fraction values of B, produces the Henry's law constant, $K_{H,A}$, for the solute A in B and $P_A = \chi_A K_{H,A}$ as in the figure above.

Finding the tangent line from a few solute P_B values allows you to determine the Henry's law constant, K_H, for B and what type of deviations occur in the binary solutions for when:

- K_H *values* $> P°$ *values* $\rightarrow$ *positive deviations are occurring*
- K_H *values* $< P°$ *values* $\rightarrow$ *negative deviations are occurring*

Benzene - toluene mixture

P-xy diagram:

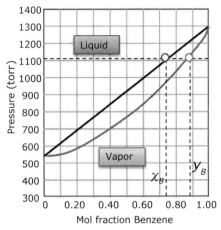

Mol fraction Benzene

KEY POINTS – LIQUID–VAPOR PHASE DIAGRAMS

If Raoult's law is obeyed, the mole fraction of B in the vapor phase cannot equal what it is in the liquid phase, so that the vapor phase over the binary solution has a different composition than in the liquid phase.

Define $y_i = $ mol fraction vapor $= P_i/P_{\text{total}}$ and $y_B = 1 - y_A$ then for a mixture of A and B:

$$y_A = \frac{P_A}{P_A + P_B} = \frac{\chi_A P_A^*}{\chi_A P_A^* + \chi_B P_B^*} = \frac{\chi_A P_A^*}{P_B^* + \chi_A(P_A^* - P_B^*)} \text{ or } y_A = \frac{P_A^*(P_{\text{total}}) - P_A^* P_B^*}{P_{\text{total}}(P_A^* - P_B^*)}$$

- In an ideal mixture, $y_A = \chi_A$ only if $P_A^* = P_B^*$
- The composition of each phase at a specific T (or P) is given by the intersection of the T (or P) with the two lines, as shown in the figures.

T-xy diagram:

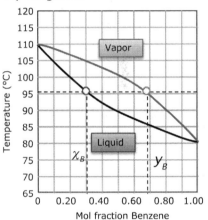

Mol fraction Benzene

The two lines that show x_i and y_i at different pressures (or T_{bp}) represent the boundaries or equilibrium lines, with the coexistence of two phases having different compositions. Only the points on the lines are meaningful and give the composition in each phase.

- The vapor phase is always "richer" in the component with the higher pure vapor pressure.
- "$P–xy$" [Pressure versus composition at a fixed T] is a plot of total vapor pressure versus the mole fraction of one component in the vapor and solution phases. These plots are useful to measure the deviation of the data from the ideal values.
- "$T–xy$" [Temperature versus composition at $P = 1.00$ atm $= 101.3$ kPa] is a plot of boiling point temperature of the mixture versus the mole fraction of one component in the vapor and solution phases. These plots are very useful for distillations and separation of phases.
- Note that the liquid phase composition is the bottom line on the $T–xy$ diagram, whereas it is the top line on the $P–xy$ diagram.
- On either plot, the mole fraction can be plotted on the x-axis as the "average" composition, z_A, where:

$$z_A = \frac{n_{A(\text{liquid})} + n_{A(\text{vapor})}}{(n_A + n_B)_{\text{liquid}} + (n_A + n_B)_{\text{vapor}}} = \frac{n_{A,\text{total}}}{(n_A + n_B)_{\text{total}}}$$

LEVER RULE LIQUID VAPOR DIAGRAMS

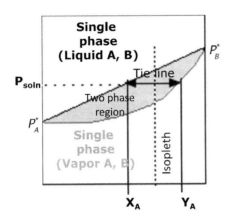

The RELATIVE amounts of each phase (liquid or vapor) at an **isopleth** (the line of constant or average composition) is given by lengths on the "tie line":

$$n_{\text{solute},\beta} \times L_\beta = n_{\text{solute},\alpha} \times L_\alpha$$

where L_α and L_β represent the lengths of the tie line from the isopleths to the equilibrium line (with the same units as mol fraction).

$$\frac{\text{Relative amt. solute in liquid phase}}{\text{Relative amt. solute in vapor phase}} = \frac{n_{B,\beta}}{n_{B,\alpha}} = \frac{L_\beta}{L_\alpha}$$

The diagrams above show that, in order for equilibrium to be maintained, the length of the segments must change with the relative amounts as the isopleth intersects at either different pressures or mole fraction.

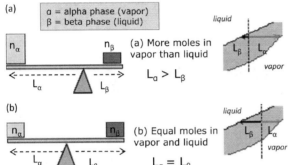

KEY POINTS – NON-IDEAL OR REGULAR SOLUTIONS: AZEOTROPES

If the solution is not ideal, the positive or negative deviations in vapor pressure produce an "azeotrope", where the mole fractions in the vapor and liquid phase are equal, and the vapor or liquid lines merge at this mole fraction. Examples of azeotropic mixtures will be given in the problems below. An ideal binary vapor diagram (like those shown above) is "zeotropic", meaning it has no azeotrope.

The observed deviations from Raoult's law should correlate to the excess enthalpy, H^E, observed for the binary solution, and create a maximum, minimum or boiling point T:

A···B forces are stronger than A···A:

- Negative deviations from Raoult's law
- Exothermic values of H^E
- The maximum boiling point is the azeotrope

A···B forces are weaker than A···A

- Positive deviations from Raoult's law
- Endothermic values of H^E
- The minimum boiling point is the azeotrope

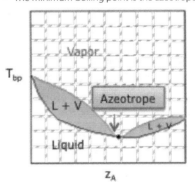

The lever rule also applies to the vapor-liquid diagram with an azeotrope, since the area between the two curves on either side of the azeotrope indicates that there are two phases present.

KEY POINTS – ACTIVITY AND ACTIVITY COEFFICIENTS FOR MIXTURES

To address the behavior of the non-ideal solutions, within the chemical potential, χ_i, is replaced with activity, a, where the activity is defined as the mole fraction times a coefficient, γ_i, so that $a_i = \gamma_i \chi_i$

$$\mu_{i,\text{soln}} = \mu_i^\circ + RT\ln\chi_i P_i^* = \mu_{i,\text{pure}}^* + RT\ln\chi_i \quad \rightarrow \quad \mu_{i,\text{soln}} = \mu_{i,\text{pure}}^* + RT\ln a_i = \mu_i^* + RT\ln(\gamma_i\chi_i) = \mu_{i,\text{ideal}}^* + RT\ln\gamma_i$$

where γ_i can never be zero but can be a decimal value or greater than 1.0. Then, the deviations observed from ideal behavior can be expressed through γ_i, where

- Positive deviations will mean $\gamma_i > 1.0$
- Negative deviations will mean $\gamma_i < 1.0$
- The activity coefficients for A and B will both be of the same type, either both greater than 1.0 or both less than 1.0

The activity coefficients may be measured many ways, but two common methods are described below:

- **Observed deviations from ideal values,** such as in vapor pressure for each component

$$\gamma_i = \frac{P_{i,\text{obs}}}{P_{i,\text{ideal}}}$$

- The G^E observed can be related to: $\ln\gamma_i$ $G_{\text{soln}}^E = \sum_i \chi_i G_i^E$ where $G_i^E = RT\ln\gamma_i$ or $\ln\gamma_i = \frac{G_i^E}{RT}$

- Since the G^E observed can be either positive or negative, this again says γ_i can be a decimal value (when $\ln\gamma_i$ is negative) or greater than 1.0 (when $\ln\gamma_i$ is positive). When $\gamma_i = 1.0$, then G^E is zero and the solution is ideal.

EXAMPLE PROBLEMS

5.17 1-Bromobutane, C_4H_9Br, has a pure vapor pressure of 28.42 mm Hg at 273 K, while that of 1-chlorobutane, C_4H_9Cl, is 1394 Pa at the same temperature.
   A) Why could we assume that this solution is an ideal solution?
   B) What would the pressure total be over a mixture of 100.0 g of each compound?
   C) What would be the mole fraction of C_4H_9Br in the vapor over the solution?
   D) What is the ΔG_{mix} for the solution?

5.18 Consider a solution containing twice as many moles of $CHCl_3$ as CCl_4, where the pure vapor pressures of $CHCl_3$ and CCl_4 are 26.54 kPa and 15.27 kPa, respectively,
   A) Why could we assume this solution is an ideal solution?
   B) What would be the composition of the vapor in equilibrium with the solution?
   C) What is the ratio of $CHCl_3$ molecules to CCl_4 in the vapor and how does this compare to the ratio in the liquid phase?
   D) For this mixture to boil at 30°C, what would the atmospheric pressure need to be?

5.19 A mixture of hexane (C_6H_{14}) and heptane (C_7H_{16}) has a total vapor pressure of 666 torr over the solution at 100°C. If the vapor pressure of pure hexane is 1836 torr at 100°C while that for pure octane is 354 torr,
   A) What is the mole fraction of hexane in the mixture?
   B) What is the mole fraction of each component in the vapor phase?
   C) (a) Prove that the ΔG of mixing of the components into the solution is negative at 100°C.
       (b) Determine how ΔG_{mix} of the solution compares to the ΔG_{mix} of the vapor phase of the mixture. Should they be the same? Explain your reasoning.

5.20 A)   Prove that equation (1) given below can be rearranged to equation (2) (below) to solve for the mole fraction of component A in the vapor of an ideal solution.

(1) $P_{total} = \dfrac{P_A^* P_B^*}{P_A^* + (P_B^* - P_A^*) y_A}$ becomes (2) $y_A = \dfrac{P_{total} P_A^* - P_A^* P_B^*}{P_{total}(P_A^* - P_B^*)}$

B)   Acetone and CCl_4 form a nearly ideal solution at 150°C where the pressure of pure acetone is 11.23 atm.

   (a)   If the total vapor pressure over the mixture is 8.76 atm when the mole fraction of acetone is 0.520, what is the vapor pressure of pure CCl_4 at the same temperature?

   (b)   Apply equation (2) to determine what the value of $y_{acetone}$ should be for the mixture.

   (c)   If the actual value is 0.631, what is the % difference between the calculated ideal value and the actual $y_{acetone}$?

   (d)   What would the activity coefficient for acetone be at $\chi_{acetone} = 0.520$?

5.21 A)   Given the graph describing the vapor pressure observed for the binary liquid mixture of A and B below:

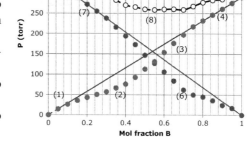

   (a)   What law has to be obeyed to produce the solid line from (1) to (5)?

   (b)   What does the line containing point (2) represent?

   (c)   Which point(s) correspond to component A acting ideally?

   (d)   Which point(s) correspond to component B acting ideally?

   (e)   Which point(s) lie in the Henry's law region for B?

   (f)   What does the line containing point (9) represent?

   (g)   What does the line containing point (8) represent?

   (h)   What does point (5) represent?

B)   Which of the following statements are TRUE for this binary mixture?

   (a)   A⋯B interactions are stronger than A⋯A or B⋯B.

   (b)   A⋯B interactions are weaker than A⋯A or B⋯B.

   (c)   A⋯B interactions are the same as A⋯A or B⋯B.

   (d)   The activity coefficients for A and B are less than 1.0.

   (e)   The activity coefficient for A is less than 1.0, but that for B is greater than 1.0.

   (f)   The activity coefficients for A and B are greater than 1.0.

   (g)   The activity coefficient for B is less than 1.0, but that for A is greater than 1.0.

   (h)   The activity coefficients for A and B are both equal to 1.0.

5.22 Given the binary mixture of A and B in the previous problem and the plot shown, answer the following questions:

A)   What can we expect to be true about the G^E and H^E of the mixture?

B)   Would an azeotrope be expected in the T–xy diagram for the mixture? Explain your reasoning.

C)   Which component has the higher value for the mole fraction in the vapor phase at $\chi_B = 0.400$?

D)   What will be true about Henry's law constant for B, $K_{H,B}$, versus P_B^*?

5.23 The data in the box on the right were collected for a binary mixture of diethyl ether and acetone at 30°C.
   A) What are the dominant forces in pure diethyl ether and acetone?
   B) Calculate P_{total} and $P_{total,ideal}$ for each mole fraction of acetone.
   C) (a) Construct a binary phase diagram for the mixture.
       (b) Is the mixture ideal? If not, what type of deviations occur?
       (c) Determine the activity coefficients for acetone and diethyl ether at $\chi_{acetone} = 0.500$

X acetone [mol/mol]	P acetone (torr)	P diethylether (torr)
0	0	646
0.05	29.4	614
0.10	58.8	581
0.20	90.3	535
0.50	168.0	391
0.80	235.0	202
1.00	283.0	0

5.24 The figure on the right shows the P–xy diagram for the diethyl ether and acetone binary mixture at three different temperatures. [Source: Dortmund Data Bank Sets 214, 632, 633]
   A) Why do the curves increase their placement on the P axis as T is increased? Do you expect this behavior to be true of all binary systems?
   B) Is the basic shape of the vapor and solution curves retained and do the lines meet at the same mole fraction at each temperature?
   C) Estimate the χ_1 (open circles) and y_1 values (filled squares) for diethyl ether at the T and P combinations given below:
       (a) $T = 303.1$ K, $P = 65$ kPa
       (b) $T = 293.15$ K, $P = 35$ kPa
       (c) $T = 273.15$ K, $P = 20$ kPa
   D) Why is the two-phase region getting smaller as T decreases?

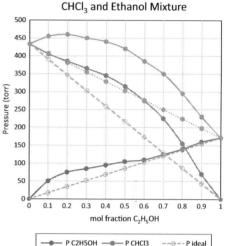

CHCl$_3$ and Ethanol Mixture

5.25 For the mixture of ethanol (C_2H_5OH) and chloroform ($CHCl_3$) at 45°C, the following data were obtained [Adapted from G. Scatchard, C.L. Raymond, JACS, (1938) 60, 1278]. Given the pure vapor pressures are 172.8 torr and 433.3 torr for ethanol and chloroform, respectively:
   A) Prepare a Raoult's law plot of the mixture by calculating the total pressure observed and, using Raoult's law, the partial pressure of each component, if ideal, and the total pressure of the ideal mixture.
   B) Concerning the completed plot:
       (a) What type of deviations are occurring for both components?
       (b) One component is much more affected, showing greater deviations from ideal behavior than the other; which component is it? What does this indicate about the A···B forces in the mixture?
       (c) How does the value of the activity coefficient for $CHCl_3$ compare to 1.0 throughout the mole fractions of C_2H_5OH?
       (d) How does the value of the activity coefficient for C_2H_5OH compare to 1.0 throughout the mole fractions of C_2H_5OH?

χ C_2H_5OH (mol/mol)	P C_2H_5OH (torr)	P $CHCl_3$ (torr)
0	0	433
0.1	50	405
0.2	75	385
0.3	85	365
0.4	95	345
0.5	105	315
0.6	110	275
0.7	125	225
0.8	140	155
0.9	160	70
1	172	0

5.26 For the binary mixture in the previous problem, both the profile of the excess functions and the log of the activity coefficients [where C_2H_5OH = "1" and $CHCl_3$ = "2"] are described by the figures below. Note that the behavior of the excess functions of H^E and S^E is more complex, but that G^E is positive through all mole fractions of ethanol. [Source: H.C. Van Ness and M.M. Abbott, Perry's Chemical Engineering Handbook (7th edition), McGraw-Hill 1997]
   A) Are the deviations we see in total pressure consistent with the G^E function? Explain the reason(s) for your decision.
   B) What can be observed in the behavior of the activity coefficient for ethanol and how does this correlate to the behavior of the H^E function of the mixture?
   C) Does the value for the activity coefficient for $CHCl_3$ shown in the second graph correlate to what was observed for the activity coefficient of $CHCl_3$ in the binary phase diagram in Problem 5.25? Explain the reason(s) for your decision.

(a)
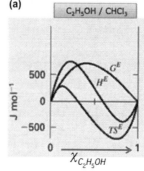

(b)
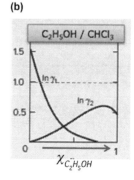

5.27 The figures on the right show the *P–xy* and *T–xy* diagrams for the ethanol–CHCl₃ system examined in the previous two problems.

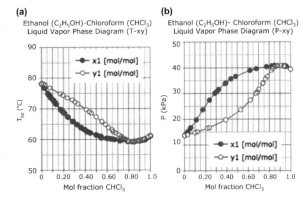

(a) Ethanol (C₂H₅OH)-Chloroform (CHCl₃) Liquid Vapor Phase Diagram (T-xy)

(b) Ethanol (C₂H₅OH)- Chloroform (CHCl₃) Liquid Vapor Phase Diagram (P-xy)

A) Does the *T–xy* plot exhibit an azeotrope? If so, is it a minimum or maximum boiling temperature? Is this expected behavior based on the observed deviations in the plot in Problem 5.25?

B) (a) Why does flipping the positions of the vapor and liquid lines cause a maximum *P* to be observed in the *P–xy* diagram rather than a minimum?

(b) Does the point where χ_1 equals y_1 stay the same in both diagrams?

C) Referring to the diagram in 5.26, do equal mole fractions in the vapor and liquid phase mean that the activity coefficients for ethanol and chloroform are the same? Explain.

D) What is the composition, in terms of mole fraction CHCl₃, of the vapor and liquid phases at 65°C?

E) If you were to choose $P = 25$ kPa, what would be the relative number of moles CHCl₃ in the vapor to the liquid phase at χ CHCl₃ = 0.4?

5.28 The *P–xy* diagram for the acetone–cyclohexane binary mixture is shown in the figure on the right. [Dortmund Data Bank Set 3849]

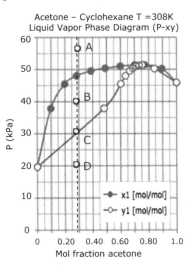

Acetone – Cyclohexane T =308K Liquid Vapor Phase Diagram (P-xy)

A) What type of deviations does this binary mixture have from ideal behavior? How does this correlate to the intermolecular forces you expect for A···B versus that in the pure components?

B) Will there be an azeotrope in the *T–xy* diagram for the mixture? If so, will it be a minimum or maximum T_{bp}? At what mole fraction of acetone would it occur?

C) For the open circles marked as points A–D, state what phases are present at each point and their respective compositions.

D) Using the lever rule, determine the relative number of moles in the liquid phase versus the vapor phase at point B.

5.29 Given the following data for an iodoethane (C₂H₅OH) and ethyl acetate (CH₃CO₂CH₂CH₃) mixture:

A) Construct the binary phase diagram showing the Raoult's law behavior and real behavior of the solution.

B) What type of forces should the A···B be? How do the A···B forces compare to A···A and B···B?

C) Estimate the Henry's law constant for using the data:

(a) For iodoethane and ethyl acetate

(b) Are the values for K_H consistent with that expected for the type of deviations observed?

D) Determine the activity coefficient values at $\chi_{iodoethane} = 0.2353, 0.5473$ and 0.8253 for:

(a) Iodoethane, C₂H₅I (b) Ethyl acetate, C₄H₈O₂, respectively.

χ C₂H₅I (mol/mol)	P C₂H₅I (torr)	P C₄H₈O₂ (torr)
0	0	280.1
0.0579	20	266.1
0.1095	52.7	252.3
0.1918	87.7	231.4
0.2353	105.4	220.8
0.3718	155.4	187.9
0.5473	213.3	144.2
0.6349	239.1	122.9
0.8253	296.9	66.6
0.9093	322.5	38.2
1.0000	353.4	0

5.30 If the boiling points of pure methanol and n-propyl bromide, C_3H_7Br, are 64.7°C and 71.0°C, respectively, and an azeotrope appears for the mixture at 54.5°C when the % by mass of C_3H_7Br, is 79.0%,
A) What type of deviations will occur in the Raoult's law plot?
B) What is the mole fraction of methanol in the azeotropic mixture?

5.31 The Raoult's law diagram for the acetone chloroform system, discussed in Problem 5.13 in terms of its excess thermodynamic properties, is shown on the right.
A) For the Raoult's law plot:
(a) Are the deviations observed consistent with what the thermodynamics had predicted? Could you tell that a complex was being formed, as we could in the plot of excess thermodynamic values, from the data in this plot? Explain the reason(s) for your choice.
(b) At what χ_{CHCl3} is the activity coefficient for $CHCl_3$ the largest negative value? Using the graph, estimate its value.
(c) Estimate the Henry's law constant for $CHCl_3$ from the graph.
B) The T–xy diagram for the acetone $CHCl_3$ system is shown on the right.
(a) Is the observed behavior consistent with the Raoult's law plot? Explain the reason(s) for your choice.
(b) What is the composition of the vapor at the azeotrope?
(c) When the mixture boils at 61.0°C what is the composition of the vapor and liquid phase?

5.32 The values for the activity coefficients are shown in the graph below for a mixture of ethanol, CH_3CH_2OH, and di-isopropyl ether (DIPE), $(CH_3)_2COC(CH_3)_2$. [Source: M.J Lee, C-H Hu, Fluid Phase Equilibria (1995) 109, 83–98]
A) Given this information, what can you expect to be true of the mixture in terms of:
(a) The actual P_{total} versus the ideal P_{total}?
(b) The value of K_H for ethanol versus $P^*_{C_2H_5OH}$?
(c) Considering the chemical structures of the two molecules, why do the A···B forces produce this type of deviation?
B) Given that the P–xy diagram for the mixture shows that the vapor and liquid phases have the same composition at $\chi_{C_2H_5OH} = 0.40$.
(a) Will there be an azeotrope in the T–xy diagram for the mixture?
(b) Will it be a maximum or minimum boiling point?
(c) At what mole fraction of ethanol will it occur?
C) Given the mole fraction for ethanol (1) and the values of ln γ for four of the mixtures below, use the equations: $G^E_{soln} = \sum_i \chi_i G^E_i$ and $G^E_i = RT \ln \gamma_i$:
(a) To calculate G^E for each component and G^E for the solution at each mole fraction.
(b) Is the sign of the G^E values for the solution the same throughout all mole fractions and is it consistent with what we expect for this mixture?

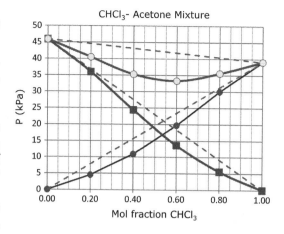

CHCl₃- Acetone Mixture

P (kPa) vs Mol fraction CHCl₃

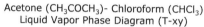

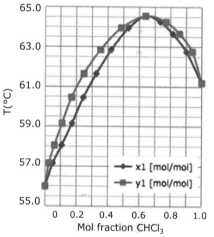

Acetone (CH_3COCH_3)- Chloroform ($CHCl_3$)
Liquid Vapor Phase Diagram (T-xy)

T(°C) vs Mol fraction CHCl₃
x1 [mol/mol]
y1 [mol/mol]

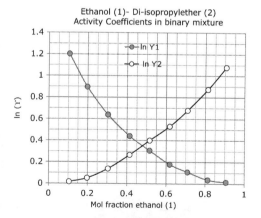

Ethanol (1)- Di-isopropylether (2)
Activity Coefficients in binary mixture

ln (Υ) vs Mol fraction ethanol (1)
ln Y1
ln Y2

x_1	ln Υ_1	ln Υ_2
0.198	0.896	0.051
0.414	0.44	0.266
0.618	0.179	0.531
0.813	0.0364	0.874

5.33 The vapor pressure of Br_2 over dilute solutions of Br_2 dissolved in CCl_4 measured at 25°C is given in the table in the box on the right.

X Br$_2$ (mol/mol)	Observed P Br$_2$ (torr)
0.00394	1.50
0.0042	1.60
0.0060	2.39
0.0102	4.27
0.0130	5.43
0.0236	9.57
0.0238	9.83
0.0250	10.27

A) Using the Clausius–Clapeyron equation, determine the vapor pressure of pure bromine at 25°C in torr. [*Look up appropriate constants.*] How would knowing this value help you determine whether the mixture is showing positive or negative deviations from Raoult's law?

B) Prove whether the mixture shows positive or negative deviations from Raoult's law. What does this tell you about the effect CCl_4 is having on the interaction in liquid Br_2?

C) Estimate the value of the Henry's law constant from the plot of the data. Is the value consistent with the type of deviations you said were occurring in the solution

5.34 The following data was measured for a binary mixture of diethyl ether ($C_4H_{10}O$) and methanol (CH_3OH) at 25°C. [Source: R. Srivastava et al., J Chem Eng Data (1988) 31, 89–93].

A) Calculate the activity coefficient for each component at each mole fraction.

B) Is the solution acting ideally? If not ideal, what type of deviations are being shown – positive or negative?

C) (a) Make a plot of log of the activity coefficients versus the mole fraction of diethyl ether (x_1).

(b) Describe any similarity to the plot in Problem 5.32? Should they be similar?

(c) Considering $\ln \gamma$ is a measure of G^E for each component (Problem 5.32), what can you say about the contribution of each to the G^E total of the mixture at $X = 0.50$?

D) Consider if the deviations prevent a solution from forming:

(a) Using the activity coefficients, G^E, for the mixture at each mole fraction.

(b) Then calculate the ΔG_{soln} from G^E for the mixture and the $\Delta G_{mix,ideal}$ at each mole fraction (see "Key Points" on page 52).

(c) Is ΔG_{soln} spontaneous at all mole fractions? When is it the least spontaneous?

Vapor Pressure Diethylether-CH$_3$OH Mixture			
		Vapor pressure @ 25°C	
1	Diethyl ether	71.54	kPa
2	Methanol	16.98	kPa
x_1	P total (kPa)	P1 (kPa)	P2 (kPa)
0.1	35.32	19.86	15.47
0.2	46.7	32.45	14.25
0.3	53.92	40.69	13.23
0.4	58.92	46.62	12.31
0.5	62.5	51.08	11.42
0.6	62.75	52.28	10.47
0.7	67.6	58.24	9.36
0.8	69.53	61.7	7.84
0.9	71.15	65.8	5.35
0.95	71.6	68.3	3.29

5.35 Given the *T–xy* diagram for a hexane and *m*-xylene mixture shown on the right [Dortmund Data Bank Set 33219] at a constant *P* of 1.00 atm. The pure vapor pressures of hexane and *m*-xylene are 1410.4 torr and 145.7 torr, respectively.

(a) Does it contain an azeotrope? What would be the name of this type of mixture?

(b) As the *T* is lowered through the open circles, marked as points A–D, state what phases are present at each point and their respective compositions.

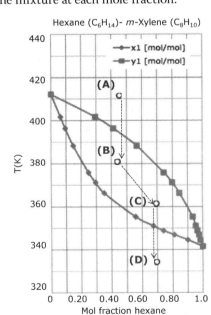

Hexane (C_6H_{14})- *m*-Xylene (C_8H_{10})

(c)   Why is the vapor always "richer" in hexane?
(d)   What would be the boiling point of a mixture that has a $\chi_{hexane} = 0.60$ in the liquid phase?
(e)   Could we determine activity coefficients for the components easily from the T–xy diagram?

KEY POINTS – COLLIGATIVE PROPERTIES

Non-volatile solutes, where the $P_B \approx 0$, produce predictable changes in the equilibrium properties of solvent A. These changes in the physical properties of solvent A are described as **colligative properties**, which depend only on the number of moles of solute particle B dissolved in the solvent A and not on the chemical identity of the solute. Thus, a colligative property equation can describe the effect for all solutes in a particular solvent.

For non-volatile solutes, Raoult's law can be rewritten to describe the decrease in the vapor pressure that will occur in the vapor pressure of solvent A over the solution, ΔP, in terms of the mole fraction of B,

$$P_A = \chi_A P_A^* = (1 - \chi_B)P_A^* \Rightarrow \Delta P = P_A^* - P_A = \chi_B P_A^*$$

FREEZING POINT DEPRESSION

$$\Delta T_{fp} = -\left[\frac{R\left(T_{fp}^*\right)^2}{\Delta \bar{H}_{fus}}\right]\chi_B = -K_{f,A}m_B$$

The lowering of the vapor pressure then changes the T at which the changes in state occur and depresses the temperature at which A freezes out of solution, $\Delta T_{fp,A} = T_{fp,soln} - T_{fp,A}^*$ (as pure A), and raises the temperature at which A boils from the solution, $\Delta T_{bp,A} = T_{bp,soln} - T_{bp,A}^*$. Both the map and your textbook explain how using the difference in the chemical potential will lead to the following general equations shown on the right (and on the map).

BOILING POINT ELEVATION

$$\Delta T_{bp} = \left[\frac{R\left(T_{bp}^*\right)^2}{\Delta \bar{H}_{vap}}\right]\chi_B = K_{b,A}m_b$$

The grouped constants, K_f, and K_b for A are called the **cryoscopic** (or *molal freezing point*) **constant** and the **ebullioscopic** (or *molal boiling point*) **constant**, respectively, and can be calculated from the appropriate phase transition temperature and enthalpy, or looked up in a reference table.

Non-volatile solutes are usually solid at room temperature and fall into one of two categories: electrolytes, which dissociate to produce ions in the solvent A, or non-electrolytes, which dissolve but do not increase the electrical conductivity of the solution. The colligative properties are used to determine how many ions an electrolyte produces per mole when dissolved in an aqueous solution. The ratio between the apparent number of ions formed and the number of moles of solute added to form the solution is called the **van't Hoff factor (i)**. Its definition is given below, as are ways it can be measured from the properties discussed so far:

$$i = \frac{\text{actual number of ions per mole solute added}}{\text{No. moles solutes added}}$$

■ The van't Hoff (i) values increase with decreasing concentration of the salt and become whole numbers (for strong electrolytes) in very dilute solutions, since ion pairing or clustering reduces the effective concentration of free ions.
■ The limiting value for the van't Hoff i factor at approximately zero concentration of solute (or infinite dilution) gives ν, the number of moles of ions per mole of salt in solution.

So, for ionic solutes, the colligative property equations become:

$$\Delta P = i(\chi_{B,added}P_A^*)$$

$$\Delta T_{mp,A} = -\left[\frac{R(T_{mp}^*)^2}{\Delta \bar{H}_{fus}}\right]\chi_B(i) = -k_{f,A}(i)\chi_{B,added}$$

$$\Delta T_{bp,A} = \left[\frac{R(T_{bp}^*)^2}{\Delta \bar{H}_{vap}}\right]\chi_B(i) = k_{b,A}(i)\chi_{B,added}$$

The true van't Hoff factor can then be calculated by comparing the actual observed property to what would have been true if no dissociation had occurred and

$$i,\text{measured} = \frac{\Delta P_{A,\text{obs}}}{\Delta P_{A,i=1.0}} = \frac{\Delta T_{fpA,\text{obs}}}{\Delta T_{fpA,i=1.0}} = \frac{\Delta T_{bpA,\text{obs}}}{\Delta T_{bpA,i=1.0}}$$

A solution of A + B separated from pure A by a semi-permeable membrane

$$\mu_A^*(P) = \mu_A(\chi_A, P + \Pi)$$

$$\downarrow$$

$$\mu_A^*(P) = \mu_A^*(P + \Pi) + RT \ln \chi_A$$

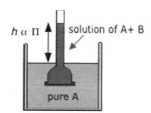

$h \alpha \Pi$ solution of A+ B

pure A

Osmosis is the most biologically important colligative property. It describes the tendency of solvent molecules A to be transported across a membrane that separates pure A from a solution of A and B where the concentration of A is lower. The pressure that has to be applied to equalize the chemical potentials and stop the flow of A across the membrane is called **the osmotic pressure**, Π. The conditions for osmosis are described in your textbook and summarized on the map.

- Osmosis is generally applied only to aqueous solutions.
- The equation then derived from the chemical potential equilibrium is: $\Pi = -\dfrac{RT}{V_A} \ln \chi_A$
- Applying some of the same assumptions as are applied to other colligative properties, as shown on map, the osmotic pressure can be defined by the **van't Hoff equation**:

$$\Pi = C_B\left(\frac{\text{mol}}{\text{L}}\right) RT = n_B\left(\frac{RT}{V_{\text{soln}}}\right)$$

The **van't Hoff factor** also applies to osmosis, if the solute is an electrolyte. The adjustment is similar in that the number of moles of solute added is multiplied by "i" so that:

$$\Pi = C_B\left(\frac{\text{mol}}{\text{L}}\right) RT(i) = (i) n_B\left(\frac{RT}{V_{\text{soln}}}\right) \quad \text{and} \quad i,\text{measured} = \frac{\Pi_{\text{obs}}}{\Pi_{i=1.0}}$$

EXAMPLE PROBLEMS

5.36 A 2.00% (w/w) solution of the non-volatile compound X in ethanol, decreases the vapor pressure by 0.45%; what is the molecular weight of the compound X?

5.37 Forty grams of $Ba(NO_3)_2$ were dissolved in 1.0 L of de-ionized water. The equilibrium vapor pressure of water was 54.909 torr over the solution at 40.0°C [$P^*_{\text{water}} = 55.324$ torr at 40°C].
    A)   What is the van't Hoff factor for $Ba(NO_3)_2$ in the solution?
    B)   What is the % dissociation of the $Ba(NO_3)_2$ in the solution?

5.38 What weight of NaCl, assuming it is 97.5% dissociated, should be added to 1.0 L of de-ionized water to lower the vapor pressure by 20.0 torr at 70.0°C, given $P^*_{\text{water}} = 233.7$ torr at 70°C.

5.39 The vapor pressure of $C_6H_6(l)$ at 60.6°C is 400 torr, but, when 19.0 g of a non-volatile organic solute was added to 500 g $C_6H_6(l)$ at 60.6°C, the vapor pressure decreased to 386 torr. What is the molecular weight of the added solute?

5.40 Sorbitol, $C_6H_{16}O_6$, is classified as a "polyol", meaning it contains more than one hydroxyl group and no carbonyl groups. It is a low-calorie sweetener in foods or soft drinks and sold commercially as a 70.0% (w/w) solution of sorbitol in water.
    A)   What is the mole fraction of sorbitol in the 70% solution?
    B)   Determine the freezing point of the 70% solution by using the:
        (a)   Equation given in the orange box on the map involving χ_B (the definition before simplifying assumptions were made).
        (b)   The commonly used equation involving the molality of the solute
    C)   Do the results of (a) and (b) agree? If not, which assumption made for the simplification of the equation in (b) may not be applicable to this solution?

5.41 If the addition of 15.0 g of a compound to 250 g of cyclohexane, $C_6H_{12}(l)$, lowered the freezing point to $-2.30°C$, given $K_f\,C_6H_{12}(l) = 20.2$ K kg/mol, $T^*_{fp} = 6.47°C$, $K_b,\,C_6H_{12}(l) = 2.79$ K kg/mol and $T^*_{bp} = 80.74°C$, what would be:
  A)   The molecular weight of the compound?
  B)   The boiling point for the solution?

5.42 What would be the expected boiling point of a solution made by dissolving 30.0 g of $I_2(s)$ in 200 mL $CHCl_3$? [$d\,CHCl_3(l) = 1.479$ kg/L, $K_b,\,CHCl_3(l) = 3.88$ K kg/mol, $T^*_{bp} = 61.2°C$]

5.43 For the colligative properties boiling point elevation or freezing point depression,
  A)   Why are K_b and K_f specific for a solvent only?
  B)   Which should be larger and why?

5.44 Ethylene carbonate (properties given in box on the right) is the solvent used for the electrolyte solution in many lithium batteries. What would be its cryoscopic constant, K_f in terms of molality in °C kg/mol?

Ethylene carbonate $C_3H_4O_3$

• Colorless organic solvent

T_{mp} 36.4 °C

ΔH_{fus} 35.8 cal/g

5.45 Diethyl ether, $C_4H_{10}O$, has a $K_f = 1.79$ K kg/mol and a normal melting point of $-114.3°C$. What is the ΔH_{fus} for diethyl ether in kJ/mol?

5.46 Assuming the solution is ideal,
  A)   Determine the boiling point of a 3.0 M solution of urea, $(NH_2)_2CO$, in water that has a density of 1.044 g/mL.
  B)   What would be the osmotic pressure of the solution?
  C)   Urea may form complexes in solution in which two or more molecules hydrogen-bond together.
    (a)   Will this effect tend to raise or lower the observed boiling point?
    (b)   Which would be more easily measured, the shift in the boiling point due to dimer formation, or osmotic pressure? Explain the reason(s) for your choice.

5.47 Although we may have some idea of what the van't Hoff factor should be for a salt from its formula, that doesn't mean we can predict its actual value in water due to ion pairing and other effects, but we can get accurate values from colligative property measurements. The table on the right gives the observed freezing point depressions for 14.0% (w/v) solutions of five salts dissolved in water [D.R.Lide, CRC Handbook of Chemistry & Physics, 88th edition, CRC Press]. For each salt, determine:

14% (by mass)	MW (g/mol)	ΔT_{fp} (°C)
NaCl	58.44	−9.94
MgSO₄	120.4	−2.86
BaCl₂	208.4	−3.92
(NH₄)₂SO₄	132.1	−4.07
AgNO₃	169.6	−2.55

  A)   The molality and van't Hoff factor for each salt and compare it to the expected value at infinite dilution.
  B)   The percent deviation from the expected infinite dilution value for each. Rank them in order, from the smallest % deviation to the largest.

5.48 Given the van't Hoff factor for the 14% $BaCl_2$ solution calculated in the previous problem, that has a density of 1.0342 g/mL,
  A)   What would be the osmotic pressure of the solution at 25°C?
  B)   By what % would you overestimate the osmotic pressure if the factor based on the formula for the salt was applied?

5.49 Applying the appropriate van't Hoff factor for the 14% (w/v) solution of $MgSO_4$ (determined in Problem 5.47), what molarity of a glucose solution would have the same osmotic pressure as the $MgSO_4$ solution?

5.50 Assume a red blood cell has an internal concentration equivalent to 0.15 M NaCl, which is fully dissociated. If the cell is to be suspended in an aqueous sucrose $(C_{12}H_{22}O_{11})$ solution,
  A)   What should be the concentration (molarity) of the sucrose so that water does not flow into or out of the cell?
  B)   Would water flow into or out of a cell if it was suspended in a 5% (w/v) dextrose in water (D5W) solution commonly used as an intravenous solution? Explain your choice.

C)   Suppose the intravenous fluid is isotonic (i.e. has same Π as cell fluid) and the original assumption of NaCl being fully dissociated was in error. What would the van't Hoff factor need to be to make the D5W solution isotonic with the cell fluid?

D)   If a 0.172 M solution of NaCl has a $\Delta T_{fp} = -0.59°C$ [D.R. Lide, CRC Handbook of Chemistry and Physics (88th edition), CRC Press], does this support using the value of van't Hoff factor calculated in (C) to obtain the answer for (A)? Explain your decision.

5.51   Mannitol, $C_6H_{14}O_6$, is the first drug of choice for the treatment of acute glaucoma in veterinary medicine. It is administered as a 20% (w/v) solution ($d = 1.06$ g/mL) intravenously. It dehydrates the vitreous humor and, therefore, lowers the intraocular pressure.
A)   What is the osmotic pressure of the 20% mannitol solution at 37°C?
B)   What is the difference between Π of the mannitol solution and that of a typical intravenous fluid, 5.0% (w/v) dextrose, $C_6H_{12}O_6$, in water?

5.52   Initially osmotic pressure is defined as $\Pi = -\dfrac{RT}{V_A} \ln\chi_A$ (from map):
A)   Derive an expression that will define the ratio between the mole fractions of water in the vitreous humor and the mannitol solution, defined in the previous problem, in terms of the difference in osmotic pressure.
B)   Using the equation, determine the actual ratio between the mole fractions of water in the vitreous humor and the mannitol solution that results in the dehydration of the vitreous humor.
C)   How does this example illustrate why the osmotic pressure is such a precise measurement of small concentration differences?

5.53   If the osmotic pressure of an unknown substance was measured at 298 K.
A)   What is the molecular weight of the substance if a 0.15% (w/v) solution of the compound in CH_3OH produced an osmotic pressure of 0.515 bar?
B)   Why does it not matter that the solvent was CH_3OH and not water, as long as we have used $\Pi V = n_B RT$ for the osmotic equation for pressure?

5.54   If 0.100 g of creatine dissolved in 1.00 L of ultrapure water produced an osmotic pressure of 13.0 torr at 25°C:
A)   What is the molar mass of creatine?
B)   What would have been the decrease in vapor pressure observed for the same solution at 25°C? Is the decrease in vapor pressure a feasible method for determining the molecular weight of creatine?
C)   If the concentration of creatine is 20 mg/L in blood and 750 mg/L in urine, what are the respective osmotic pressure values at 25°C for blood and urine? Could they be easily measured?

5.55   Consider a water-soluble polymer that has an average molar mass of 210,000 g/mol.
A)   What molarity would be needed to produce an osmotic pressure of 10 torr at 37°C?
B)   What weight in grams of polymer would need to be dissolved per kilogram of water to achieve this pressure?

5.56   Polymer molecules often aggregate, even in dilute solutions, so a consistent average molecular weight is difficult to measure. However, if π values are measured as a function of polymer concentration in g/L, an accurate average molar mass can be determined.

T = 44°C					
Osmotic Pressure (Pa)	15	30	50	70	85
Conc. (g/L)	2	5	7	10	12.5

A)   Prove that plotting π versus dissolved polymer concentration in g/L can give you the molar mass of the polymer at infinite dilution.
B)   Apply the plotting method to the data in the table (converting to convenient units as needed) to get the average molar mass of polystyrene in solution with cyclohexane.

5.57 Apply the same plotting method in the previous problem to the data below (converting to convenient units as needed) to get the average molar mass of bovine serum albumin.

T=37°C				
Osmotic Pressure (torr)	2.51	5.07	8.35	19.33
Conc. (g/L)	8.95	17.69	27.38	56.20

5.58 A 0.300 m $C_6H_{12}O_6$ solution has a density of 1.038 g/mL (density of water = 0.9974 g/mL at 25°C, if needed).
  A) Calculate the osmotic pressure at 25°C using the three equations (a)–(c) derived by making the approximations described on the map (and in your textbook):

  (a) $\Pi = -\dfrac{RT}{V_A}\ln\chi_A$ (b) $\Pi = \dfrac{RT}{V_A}\chi_B$ (c) $\Pi V = n_B RT$ (V = volume solution)

  B) Comparing the values for Π, which approximation(s) produce the most significant change?

5.59 The osmotic pressure of a compound dissolved in cyclohexane at 288 K is 99.0 kPa. What is the freezing point of the solution? [Constants for C_6H_{12}(l) in Problem 5.41, d C_6H_{12}(l) = 779 kg/m³]

5.60 A solution that is 0.2164 M mannitol ($d = 1.005$ g/mL) has the same osmotic pressure as a 0.165 M NaCl solution ($d = 1.012$ g/mL).
  A) What is the freezing point of the NaCl solution?
  B) If the two solutions have the same osmotic pressure, will they also have the same freezing point depression? Explain your decision.

5.61 Human blood plasma contains approximately 40.0 g of albumin (MW = 69,000) and 20.0 g of globulin (MW = 160,000) per liter.
  A) Calculate the osmotic pressure due to these two components at 37°C.
  B) How many times greater is the osmotic pressure due the ions in blood which is the same as that resulting from a 5% (w/v) dextrose ($C_6H_{12}O_6$) solution in water?
  C) Why could we assume that the osmotic pressures from the different solutes will add together and not cancel each other out?

5.62 Reverse osmosis requires the application of a pressure greater than Π to force pure water from the solution side across a suitable membrane to be collected as pure water. If seawater has an average total concentration of 1.0 M solute ions at 300 K what would be the minimum pressure, in atm, needed to reverse the flow of water?

5.63 Brackish water has a higher salt content than freshwater, but below that of seawater. If a sample of brackish water has the concentration, in ppm, of the six major ions given in the table below, what would be the minimum pressure, in atm, needed to reverse the flow of water at 25°C and obtain pure water from the brackish water?

Ion	Na^+	Mg^{+2}	Ca^{+2}	Cl^-	SO_4^{-2}	HCO_3^-
Concentration (ppm)	1837	130	85	2970	479	250

KEY POINTS – ACTIVITY COEFFICIENTS IN COLLIGATIVE PROPERTIES FOR REAL SOLUTIONS

Activity and activity coefficients for solvent A may also play a role in colligative properties, causing the observed property to be either higher or lower than the expected ideal behavior. The original equations are adjusted by defining the activity as: $a_A = \gamma_A\chi_A$ so that $\ln a_A$ replaces $\ln \chi_A$ and:

$$\Delta T_{fp} = \left[\frac{R(T_{fp}^*)^2}{\Delta H_{fus,A}}\right]\ln a_A \qquad \Delta T_{bp} = -\left[\frac{R(T_{bp}^*)^2}{\Delta H_{vap,A}}\right]\ln a_A \qquad \Pi = \left[\frac{RT}{V_A}\right]\ln a_A$$

which leads to: $\Delta T_{fp} = -K_{f,A}\gamma_A m_B$ $\Delta T_{bp} = K_{b,A}\gamma_A m_B$ $\Pi = \gamma_A C_B(mol/L)RT$

As before, the activity coefficient can be measured by comparing the actual, observed colligative property to that expected when the activity coefficient is 1.0.

• The γ_A for **vapor pressure** P_A would be defined as: $\gamma_A(\text{non-electrolyte } B) = \dfrac{P_{A,obs}}{\chi_A P_A^*}$	• The effect on γ_A for **non-electrolyte solutes** can be found by comparing the observed property to the value when $\gamma_A = 1.0$: $\gamma_A = \dfrac{\Delta T_{fp,obs}}{\Delta T_{fp,\gamma=1.0}}$ $\gamma_A = \dfrac{\Delta T_{bp,obs}}{\Delta T_{bp,\gamma=1.0}}$ $\gamma_A = \dfrac{\Pi_{obs}}{\Pi_{\gamma=1.0}}$	• The effect on γ_A for **electrolyte solute** can be found only after determining the correct value for the van't Hoff factor of the solute.

EXAMPLE PROBLEMS

5.64 At 310 K, the vapor pressure of a 60.0% (w/v) solution of glycerol, $C_3H_8O_3$, in water is 33.98 torr. If the vapor pressure of pure water is 47.12 torr at 310 K,
  A) What is the activity coefficient for water in the solution at this temperature?
  B) The activity of water in foods above 0.80 can slow down enzymatic reaction or enhance metal oxidation reactions if above 0.30. Would the activity of water in this solution exceed either of these limits?

5.65 Raffinose, $C_{18}H_{32}O_{16}$, is an abundant trisaccharide found naturally in many foods.
  A) Dilute solutions of raffinose showed the vapor pressures at 318 K given in the table on the right.

	X raffinose	P H$_2$O (kPa)
H$_2$O(l)	0	9.514
Solution I	0.02099	9.239
Solution II	0.04149	8.732

  (a) What are the values of the activity coefficients for water in the raffinose solutions I and II?
  (b) Does the activity coefficient stay the same? Should we expect it to stay constant? Explain the reason(s) for your choice.
  B) Given the values, what sign can we expect for H^E of mixing for the solutions and what does this tell us about the strength of the A···B interactions in the solution?
  C) If the maximum solubility of raffinose in water is 203 kg/m³ of water, what would be the:
  (a) Expected freezing point for the solution if the solution were ideal?
  (b) Given the behavior observed in the solutions in A), what should we expect to be true about the activity of water in the saturated solution and how would this affect the observed ΔT_{fp}?

5.66 The data obtained for sorbitol solutions indicate that the measured vapor pressure over the 70% sorbitol solution described in Problem 5.40 is 6.877 kPa (51.6 torr) at 45°C. What would be the:
  A) Activity coefficient of the water in the 70% solution from the vapor pressure?
  B) Freezing point depression, adjusted for the activity coefficient of water calculated in A), for the 70% solution, using the result from 5.40 B) (b) as the ideal solution value?
  Furthermore
  C) How much of a difference does including activity make in the expected ΔT_{fp}?

5.67 A solution is prepared by adding 105.56 g of mannitol, $C_6H_{14}O_6$ [formula weight, FW = 182.2], to 500 mL of water at 35°C, when $P_{H_2O}^* = 42.2$ torr at 35°C. If the observed vapor pressure is 40.51 torr, what is the:
  A) Activity coefficient for water in this solution?
  B) Freezing point of the solution, if you take the activity coefficient into account?
  C) Osmotic pressure of the solution at 35°C if you take the activity coefficient into account? [density of the solution = 1.06 g/mL]
  Furthermore
  D) Did applying the activity coefficient for water make a significant difference to the osmotic pressure? (Explain your answer.)

5.68 If a 60% (w/v) glycerol, $CH_2(OH)CH(OH)CH_2(OH)$, solution in water has a measured vapor pressure of 0.434 atm while pure water has a vapor pressure of 0.620 atm at the same temperature:
  A) What is the activity coefficient for water in this solution?
  B) What is the ratio of water molecules to glycerol molecules in this solution? Would you have expected the solution to be an ideal solution? What could you expect about the interactions between glycerol and water in this solution?
  C) When a solution has fewer "free" water molecules, because they are tightly bound to the solute, this effectively lowers the χ_A (since there are not as many independent molecules) which has the effect of appearing to increase χ_B above its actual value. Is that the case in this solution? Explain your decision.

5.69 A solution is made by dissolving 80.0 g of lactose, $C_{12}H_{22}O_{11}$, in 920 g of water.
  A) Determine the value of γ_{H_2O} if the published value for the freezing point depression of the solution is $-0.50°C$. [D.R. Lide (ed.), CRC Handbook of Chemistry and Physics, 88th edition, CRC Press].
  B) What is different about this value of γ_{H_2O} as opposed to the values seen in earlier problems?
  C) Before, in the binary solutions made from two volatile components, we saw activity coefficients that were either less than 1.0 or greater than one, dependent on the strength of the A···B interactions. Are we seeing the same possibilities with non-volatile solutes added to a solvent? Explain the reason(s) for your decision.

5.70 We know that water molecules in pure water are clustered into H-bonded groups of molecules and not "free", unbound water molecules. The activity coefficient can be seen as a measure of how many "free" molecules exist in solution versus that in pure water. Given the properties of 20.0% (w/v) solutions of ethylene glycol, 1-propanol, or urea in the table below:

Solution: 20%(w/w)	Solute:	molality (m)	Molarity (M)	ΔT_{fp} (°C)
Ethylene glycol	$HOCH_2CH_2OH$	4.028	3.3	−7.93
1-propanol	$CH_3CH_2CH_2OH$	4.16	3.223	−7.76
Urea	H_2NCONH_2	4.163	3.506	−7.00

  A) What is the activity coefficient for water in each solution based on the published freezing point depression [D.R. Lide (ed.), CRC Handbook of Chemistry and Physics, 88th edition, CRC Press]?
  B) Correlate the observed activity coefficient of water to the solute···water (A···B) interactions and state whether the relative values of "free" water in the solution have increased, decreased or remained about the same as in pure water.
  C) Apply the appropriate value of the activity coefficient to calculate the osmotic pressure at 25°C for the urea solution. By what % would you overestimate the osmotic pressure at 25°C if the activity coefficient were not applied? Is this a significant difference?

KEY POINTS – HENRY'S LAW AND SATURATED SOLUTIONS OF SOLUTE B IN A

(Solubility as a Colligative Property Equation)

Henry's law for binary solutions, defined earlier, can be used to describe the saturation concentration of a dissolved gas, B, in a solvent A, since the Henry's law constant best defines the mole fraction in a dilute solution.

▪ Combining $\chi_B = \dfrac{P_{gas\,B}}{K_{H,B}}$ and $\chi_B = \dfrac{n_{gas\,B}}{n_A + n_{gas\,B}} \approx \dfrac{n_{gas\,B}}{n_A}$ leads to: $\dfrac{P_{gas\,B}}{K_{H,B}} = \dfrac{n_{gas\,B}}{n_A}$

- Then defining n_A: $\dfrac{n_{gas\,B}}{V_{soln}} = \dfrac{P_{gas\,B}}{K_{H,B}V_A}$ produces **Henry's law for gases**: $C_B\left(\dfrac{mol\,B}{L}\right) = \dfrac{P_{gas\,B}}{K_{H,B}}$

- Note that the units on $K_{H,B}$ have become atm L/mol instead of just pressure units quoted for binary liquid mixtures and that it changes value with different solvents.
- The $K_{H,B}$ values are typically measured for saturated solutions of the gas at different temperatures in the solvent and tabled, rather than converting from the binary solution values.
- The Henry's law equation for gases can appear in either of two ways in the literature:

$$P_{gas\,B} = K_{H,B}C_B\left(\frac{mol}{L}\right) \quad \text{or as:} \quad C_B\left(\frac{mol}{L}\right) = K_{H,B}P_{gas\,B}$$

- The units given for tabled values of $K_{H,B}$ indicate which equation is being used.

For solutes other than gases, the solubility of a non-electrolyte solute B in an ideal saturated solution of A and B can be defined by considering the equilibrium between the chemical potentials $\mu_B(s)^* = \mu_B(soln)$ in the saturated solution. Using a similar derivation pathway as used for the colligative properties, the

solubility of a non-gaseous solute B in a saturated solution can then be defined at any temperature, T, below the melting point of the solute, by:

$$\ln\chi_B = \frac{\Delta H_{fus,B}}{R}\left[\frac{1}{T^*_{mp,B}} - \frac{1}{T}\right]$$

Because the value of χ_B depends on the properties of B, not A, and on the chemical identity of the solute, solubility is not like other colligative properties which depend only on the properties of A. However, since the properties of only one component have to be considered, the solubility equation is often included with the colligatve properties. If the solution is not ideal, however, this equation for $\ln \chi_B$ does not apply, so it has limited use.

EXAMPLE PROBLEMS

5.71 Given the Henry's law constant for Argon gas in water is given in a reference table as 1.4×10^{-5} mol/m³ Pa what would the pressure of Ar(g) needed to produce a concentration in water of the Ar(g) that was 0.010 M?

5.72 The mole fraction of CCl_2F_2 gas in a saturated aqueous solution was 4.17×10^{-5} at 1.0 atm pressure of the gas. What is the:
   A) Approximate molarity of the saturated solution?
   B) Henry's law constant for CCl_2F_2 gas in water?

5.73 Henry's law constant plays a fundamental role in developing accurate models for the transport of volatile or semi-volatile organic pollutants through water, oceans and the atmosphere, that may persist in the environment. Of particular concern are pesticides that are harmful when ingested or inhaled since water sources or the atmosphere can then be routes of exposure. Henry's law constant for heptachlor, a chloro-organic pesticide, has been determined in both de-ionized water and saline solution and found to be 82.6 and 211.6 Pa m³/mol, respectively, at 35°C. If the maximum solubility for heptachlor is 0.056 mg/L in water,

Heptachlor
MW = 393.35

A)   What would be the pressure of heptochlor above the water solution in atm?

B)   For practical use, the partial pressures of gaseous pollutants are converted to concentrations of molecules/cm³, using the ideal gas law. What would be the number of molecules of heptachlor per cm³ above the saturated solution?

C)   Does the higher K_H in saline solution mean heptachlor is less likely or more likely to remain in seawater, once dissolved?

D)   Assuming the same solubility, what would the concentration of heptachlor be in molecules/cm³ above the saline solution (as a substitute for seawater)?

5.74 If the ΔH_{fus} for phenanthrene, $C_{14}H_{10}$, is 4450 cal/mol and its melting point is 100°C,

A)   What is the molar solubility of phenanthrene in benzene at 20°C, if they form an ideal solution?

B)   If the literature value given for a saturated solution of phenanthrene in benzene at 20°C is 1.0 g in 2.0 mL benzene, does your value in A) agree with this value? Explain your decision [density $C_6H_6 = 876$ kg/m³, density $C_{14}H_{10} = 1179$ kg/m³].

5.75 For a solution of p-dichlorobenzene, $C_6H_4Cl_2$, in hexane, C_6H_{14}:

A)   Determine the expected solubility at 20°C, as a mole fraction, of p-dichlorobenzene in hexane when the melting point of $C_6H_4Cl_2$ is 52.7°C and its $\Delta H_{fus} = 18.16$ kJ/mol.

B)   If the published value for a saturated solution is 20.57% (w/v) $C_6H_4Cl_2$ in hexane, is the $C_6H_4Cl_2$–C_6H_6 solution behaving ideally?

C)   If not ideal, what could be the reason(s) why it's not acting as an ideal solution?

5.76 Naphthalene, $C_{10}H_8$, forms an ideal solution with liquid benzene, C_6H_6. At what T in °C would there be 1.0 g naphthalene dissolved per g benzene?

Free Energy, Equilibrium Constants, and Electromotive Force

6

KEY POINTS – FREE ENERGY AND CHEMICAL REACTIONS

In Part 4 of this guide, we introduced Gibbs free energy (ΔG) and its importance in determining the direction of a chemical reaction. Part 5 illustrated how defining Gibbs free energy as the chemical potential would allow the definition of many solution properties and that equilibrium occurs when the difference in the chemical potentials between two states is zero. In Part 6, we will extend the application of the chemical potential and ΔG to equilibrium in chemical reaction systems, as opposed to pure substances or the non-reacting mixtures previously discussed in Part 5.

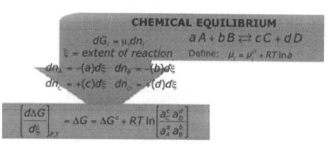

The standard state value, $\Delta\mu°$ or $\Delta G°$, for a substance is measured at 25°C and 1.0 atm, and concentration = 1.0 M, if in solution, and used to determine the spontaneity of the reaction given:
$\Delta G° = \Delta H° - T\Delta S°$

The free energy for the reaction also sets the value of K, the ratio of the activities for the products over the reactants and the extent of reaction, $\Delta G_r^° = -RT \ln K$.

- If $\Delta G°$ (−) spontaneous reaction, $K > 1.0 \rightarrow$ reaction is "strong", meaning that there are more products than reactants at equilibrium.
- If $\Delta G°$ (+) non-spontaneous reaction, $K < 1.0 \rightarrow$ reaction is "weak", meaning that there are more reactants than products at equilibrium.

All chemical reactions have both a forward rate of reaction and a reverse rate, so that a dynamic equilibrium state, of constant composition, is achieved when the forward rate equals the reverse rate of reaction. The extent of reaction needed to achieve that balance then defines K, the equilibrium constant, either in pressures (if a gaseous reaction) or mole fractions or molarities, if in the solution phase.

- K must be a dimensionless ratio, so the unit must be defined.
- $K_P = \dfrac{\left(P_C/P°\right)^c \left(P_D/P°\right)^d}{\left(P_A/P°\right)^a \left(P_B/P°\right)^b}$ where $P° = 1.0$ atm or $K_C = \dfrac{\left([C]/1.0\,\mathrm{M}\right)^c \left([D]/1.0\,\mathrm{M}\right)^d}{\left([A]/1.0\,\mathrm{M}\right)^a \left([B]/1.0\,\mathrm{M}\right)^b}$ where $[X] =$ molarity
- $K_p = K_c\left(RT\right)^{\Delta n}$ where Δn is the same as defined earlier in Part 2.

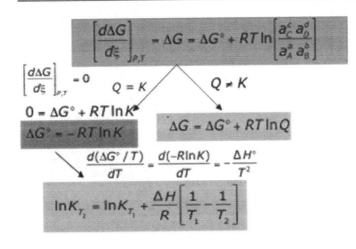

Although, in chemical reaction systems, the $\Delta G°$ value for the reaction ($\Delta G_r°$) cannot be zero, the ΔG_r value (the non-standard value), derived from the chemical potential changes, is zero at equilibrium (as explained in your textbook and illustrated on the map). This then leads to the difference between K, the equilibrium constant, a fixed value for the products and reactants, and the reaction quotient, Q, which is not fixed and reflects non-equilibrium conditions.

- When $Q = K$ the system is at equilibrium
- When $Q < K$ the conversion of reactants to products is spontaneous, and the forward rate is greater than the reverse rate of reaction
- When $Q > K$ the conversion of products to reactants is spontaneous, and the reverse rate is greater than the forward rate of reaction.

The van't Hoff equation defines both the connection of K to the $\Delta H°$ and $\Delta S°$ for a reaction, as well the value of $\ln K$ at different temperatures as shown above (and on the map), assuming $\Delta S°$ is constant.

Van't Hoff equation: $\left[\dfrac{\partial(\ln K)}{\partial T}\right]_P = -\dfrac{\Delta H_r°}{RT^2} \Rightarrow \ln K = -\dfrac{\Delta H_r°}{RT} + \dfrac{\Delta S_r°}{R}$

The van't Hoff equation indicates that the sign of ΔH for the reaction determines how K will change with T, since, assuming $\Delta S_r°$ is constant $T_1 \rightarrow T_2$, then:

- When ΔH_r (+), K and T are directly related so that $K \uparrow$ when $T \uparrow$ whereas $K \downarrow$ when $T \downarrow$.
- When ΔH_r (–), K and T are inversely related so that $K \downarrow$ when $T \uparrow$ whereas $K \uparrow$ when $T \downarrow$.

Since K can only change value if T changes, pressure changes cannot change the value of K_p. However, if the equilibrium system involves gases, the equilibrium composition can be altered when the total pressure changes to keep the partial pressure ratio defined by K. This is summarized by Le Chatelier's principle, that says that, if the system volume contracts, the reaction rate favoring the side of the with the least number of moles of gas will dominate and change the percent dissociation. The opposite will happen if the volume expands.

Coupled reactions play a crucial role in biological systems but are also important for other, chemical applications. Coupling exergonic ($\Delta G°$ (–)) and endergonic ($\Delta G°$ (+)) reactions will increase the extent of the reaction for the endergonic reaction. The end result of coupling reactions is that the K's for each individual reaction will be multiplied together to produce K for the overall reaction:

$$K_{overall} = K_a K_b K_c$$

COUPLING REACTIONS:

Addition of known reactions

a(reaction 1)	$a \times \Delta G_1$
+b(reaction 2)	$b \times \Delta G_2$
+c(reaction 3)	$c \times \Delta G_3$

$\Delta G_{total}° = a\Delta G_1° + b\Delta G_2° + c\Delta G_3°$.

EXAMPLE PROBLEMS

6.1   Ammonium carbamate, $(NH_3)_2CO_2(s)$, dissociates to produce $NH_3(g)$ and $CO_2(g)$: $(NH_3)_2CO_2(s) \Leftrightarrow 2NH_3(g) + CO_2(g)$. At 30°C, the dissociation pressure (P_{total}) was found to be 124 mmHg.

    A)   What is the K_p for the reaction at 30°C?

    B)   Calculate the $\Delta G_f°$ for $(NH_4)_2CO_3(s)$ from the K_p for the reaction, assuming that the $\Delta G_{298}°$ values for $NH_3(g)$ and $HCl(g)$ remain constant.

6.2   The dissociation of solid silver oxide occurs as: $2\,Ag_2O(s) \Leftrightarrow 4\,Ag(s) + O_2(g)$ at 25°C

    A)   For the balanced reaction:

       (a)   What is the $\Delta G°$ for the reaction at 25°C?

(b)   How would K_p be defined for this reaction?

(c)   Should K be greater or less than 1.0 for this reaction at 25°C?

B)   Calculate the pressure of $O_2(g)$ present at equilibrium for the dissociation at 25°C.

C)   Determine the T (°C) at which the pressure of $O_2(g)$ will be 1.0 atm.

6.3 $NO_2(g)$ undergoes dissociation to produce $NO(g)$ and $O_2(g)$. If one mole of $NO_2(g)$ is placed in a container and allowed to come to equilibrium at a total pressure of 1.0 atm at 700 K and then 800 K, the following results will be obtained:

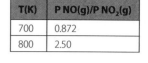

T(K)	P NO(g)/P NO$_2$(g)
700	0.872
800	2.50

A)   Write the balanced reaction for the dissociation of 1.0 mole of $NO_2(g)$ and the proper expression for K_p.

B)   Calculate the values of K_p at 700 and 800 K and then the values of $\Delta G°$ for the reaction from the experimental data.

C)   What do the results in B) tell you about the signs of $\Delta H°$ and $\Delta S°$ for the reaction?

D)   Calculate the values of $\Delta H°$ and $\Delta S°$ for the reaction, using the tabled 298 K values, and then calculate the "changeover" T for the sign in $\Delta G°$. Do the results correlate to what was observed in the experiment?

6.4   For the reaction: $H_2(g) + I_2(g) \Leftrightarrow 2\ HI(g)$, the partial pressures of $H_2(g)$, $I_2(g)$ and $HI(g)$ at equilibrium were measured at 731 K and are given on the table on the right.

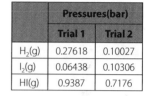

	Pressures(bar)	
	Trial 1	Trial 2
H$_2$(g)	0.27618	0.10027
I$_2$(g)	0.06438	0.10306
HI(g)	0.9387	0.7176

A)   Using the data,

(a)   Determine the average K_p for the reaction with the pressures given.

(b)   Does it make a difference in the value of K_p that the pressures are given in bars, not atmospheres? Would that always be true?

B)   Calculate $\Delta G°$ for the reaction at 731 K from the equilibrium constant.

C)   Given that ΔC_p is –7.10 J/K for the reaction, calculate the value of $\Delta G_r°$ at 731 K and compare it to the value determined in B). Do they agree?

6.5 PCl_5 has a melting point of 160.5°C and a boiling point of 166.8°C. As a gas, it decomposes according to the reaction: $PCl_5(g) \Leftrightarrow PCl_3(g) + Cl_2(g)$, and has a $K_p = 1.05$ at 250°C. If you start with 5.0 g $PCl_5(s)$ in an evacuated 2.0-L container, heat it to 250°C, and allow the equilibrium to occur,

A)   What pressure of PCl_5 gas will be observed in the container at 250°C?

B)   (a)   What is the % of PCl_5 gas in the equilibrium mixture at 250°C?

(b)   Do the relative amounts of product to reactant correlate to what is expected when $K > 1.0$? Explain the reason(s) for your decision.

C)   If cooled to 20°C, the $PCl_5(g)$ should solidify. What would be the weight of $PCl_5(s)$ left in the container at 20°C?

6.6   Borneal, $C_{10}H_{18}O$, a terpene used in perfumes, has the structure and properties given in the box on the right. Borneal isomerizes to iso-borneal with a $K_c = 0.106$ at 230°C. If 16.0 g of borneal powder is combined with 4.0 g of iso-borneal in a 2.0-L container, then heated to 230°C and allowed to come to equilibrium:

A)   Will the concentrations remain the same? If not, which way will they shift?

B)   After equilibrium has been established, what will be the weights of each isomer in the mixture?

C)   What is the value of ΔG for the initial reaction system?

Borneol

Isoborneol

m.p. 208°C   sublimes
b.p. 212°C   at 214°C

6.7   Amylene, C_5H_{10}, and acetic acid, CH_3CO_2H, react to give the ester, amyl acetate, which has an odor like that of pears. If 6.45×10^{-3} mole of amylene is combined with 1.00×10^{-3} mole of acetic acid in 845 mL of an inert solvent and 7.84×10^{-4} moles of the ester was formed at 25°C, what was the:

Acetic acid   +   Amylene $\Leftrightarrow$ Amylacetate

A) K_c for the reaction at 25°C?

B)   % conversion based on the limiting reactant? Does this % correlate to the value of K_c?

C) $\Delta G°$ for the reaction at 25°C?

6.8   Solid HgO dissociates according to the reaction: $2HgO(s) \Leftrightarrow 2\,Hg(g) + O_2(g)$. If the total pressure is 50.0 kPa at 420°C in the equilibrium system and 108 kPa at 450°C in the same container, what are:

A)   The equilibrium constants for the two temperatures, as K_p values?

B)   The equilibrium constants for the two temperatures, as K_C values?

Furthermore:

C)   Will the enthalpy of dissociation for HgO(s) calculated from the K values be different if you use the K_c values instead of the K_p values? Explain your choice.

D)   What is the approximate value for the ΔH_r in the T range, given the values calculated in C)?

6.9   One way to check the accuracy of the thermodynamic values, for ions and other aqueous species, is to calculate K_c using the $\Delta G°$ value and compare the result to known values. For the weak base, NH_3, the dissociation reaction can be described as: $NH_3(aq) + H_2O(l) \Leftrightarrow NH_4^+(aq) + OH^-(aq)$ with $K_b = 1.76 \times 10^{-5}$ at 25°C. Using the thermodynamic values from standard tables, what is the:

A) $\Delta G°$ for the reaction?

B) K_c for the reaction at 25°C? Compare the calculated K to the K_b value given in the problem.

C)   Will the spontaneity of the reaction change sign when T is changed? Explain the reason(s) for your decision.

6.10 Copper (II) chloride dihydrate dehydrates to form copper (II) chloride monohydrate and water vapor

$$CuCl_2 \cdot 2H_2O(s) \Leftrightarrow CuCl_2 \cdot H_2O(s) + H_2O(g)$$

T (°C)	P (torr)
17.9	0.0049
39.8	0.025
60	0.122
80	0.327

Given the data in the table on the right, determine:

A)   The log of the equilibrium constant, $\ln K_p$, at each temperature.

B)   The enthalpy for the dehydration from the data

C)   The values of $\Delta G°$ and $\Delta S°$ at 20.0°C

6.11 For the reaction: $I_2(g) + cyclopentane(g) \Leftrightarrow 2HI(g) + cyclopentadiene(g)$

A)   Given that the temperature dependence follows the equation: $\log K_p = 7.55 -$, $4844 \div (T(K))$ calculate:

(a)   The K_p and $\Delta G°$ at 400°C for the reaction

(b)   The $\Delta H°$ and $\Delta S°$ at 400°C for the reaction

B)   Calculate the K_c from the K_p at 400°C.

C)   If the starting concentrations of both reactants are 0.10 M, estimate the concentration of HI at equilibrium at 400°C in the system.

6.12 Under standard-state conditions, one of the steps in glycolysis does not occur spontaneously.

$$Glucose + HPO_4^{-2} \rightarrow glucose\text{-}6\text{-}phosphate + H_2O(l)\quad \Delta G°' = 13.4\,kJ$$

Could the reaction be spontaneous in the cytoplasm of the cell where the concentrations are [glucose] = 0.045 M, $[HPO_4^{-2}]$ = 0.0027 M, [glucose-6-phosphate] = 1.6×10^{-4} M at 37°C? Explain your decision.

6.13 Given the following two reactions, where PEP = phosphoenolpyruvic acid, Py = pyruvic acid, P = phosphate ion, ADP = adenosine diphosphate, and ATP = adenosine triphosphate

$$PEP + H_2O(l) \rightarrow Py + P\qquad \Delta G° = -53.6\,kJ$$
$$ADP + P \rightarrow ATP + H_2O(l)\qquad \Delta G° = +29.3\,kJ$$

A)   Prove that coupling the reactions will produce the spontaneous conversion of ADP to ATP and calculate K for the combined reaction.

B)   (a)   Determine the concentration of ATP produced at 298 K by the equilibrium if the initial concentrations of PEP and ADP are 0.01 M.

       (b)   What is the percent conversion of ADP to ATP (and PEP to pyruvate) in the coupled reaction at 298 K?

       C)   Determine what the K would have been for the first reaction *without* the coupling. How is the percent conversion affected by the coupling?

6.14   The following reactions can be coupled to give alanine and oxaloacetate at 30°C:

$$\text{glutamate} + \text{pyruvate} \rightarrow \text{ketoglutarate} + \text{alanine} \qquad \Delta G° = -1004 \text{ J}$$
$$\text{glutamate} + \text{oxaloacetate} \rightarrow \text{ketoglutarate} + \text{aspartate} \qquad \Delta G° = -4812 \text{ J}$$

A)   Write the form of the equilibrium constant for the coupled reaction, shown below, and calculate the numerical value of the K for the coupled reaction at 30°C.

$$\text{pyruvate} + \text{aspartate} \rightarrow \text{alanine} + \text{oxaloacetate}$$

B)   In the cytoplasm of a certain cell, the components are at the following concentrations: pyruvate $= 10^{-2}$ M, aspartate $= 10^{-2}$ M, alanine $= 10^{-4}$ M, and oxaloacetate $= 10^{-5}$ M. Calculate the Gibbs free-energy change for the added reaction under these conditions. Will the reaction be spontaneous under these conditions?

6.15   Flavin adenine dinucleotide (FAD) and reduced nicotinamide adenine dinucleotide (NADH) are coupled in the reaction sequence described in the figure below. Given that the free-energy changes for the individual reactions are:

$$\text{FAD} + 2\text{H}^+ + 2\text{e}^- \rightarrow \text{FADH}_2 \qquad \Delta G°{'} = 42.3 \text{ kJ}$$
$$\text{NAD}^+ + \text{H}^+ + 2\text{e}^+ \rightarrow \text{NADPH} \qquad \Delta G°{'} = 65.5 \text{ kJ}$$

A)   For the added reactions being depicted in the figure on the right, write the balanced overall reaction and determine the $\Delta G°{'}$ for the overall reaction.

B)   Determine the K_c for the overall reaction.

C)   The ΔG values are those for a solution with pH $= 7.0$. Does pH influence the value of K in this coupled reaction? Why is it important that the pH be known in order to quote the correct value for K for biochemical reactions like this one?

6.16   As part of the citric acid cycle, the citrate ion is first converted to *cis*-aconitate and water, as shown in the figure on the right as STEP I. Given that the $\Delta G_f°$ values for citrate ion(aq) and *cis*-aconitate(aq) are -1167.1 and -921.7 kJ/mol, respectively:

A)   What is the $\Delta G°$ for the STEP I reaction?

B)   What concentration of citrate would be in equilibrium with 12.0 mM *cis*-aconitate at 25°C as a result of STEP I?

C)   In the next step (STEP II), the aconitate and H_2O are converted to isocitrate ion and the overall two-step reaction has a $\Delta G°$ of 13.3 kJ/mol. What is the $\Delta G°$ for STEP II?

D)   Using the overall K_c for the sequence, what is the K_c for the second step?

6.17   The conversion of 3-phosphoglycerate [3-Pgly] ⇔ 2-phosphoglycerate [2-Pgly] occurs at pH 7 and shows the K_c values at various temperatures given in the table below.

T (°C)	K
0.0	0.1538
20.0	0.1558
30.0	0.1569
45.0	0.1584

A)   Determine the $\Delta H°$ for the reaction at pH = 7.0

B)   Calculate $\Delta G°$ for the reaction at 25°C and a pH = 7.0

C)   What is the equilibrium concentration of each isomer when 0.150 mmol 3-Pgly is initially added to 1.0 mL of water at 25°C?

KEY POINTS – IONIC STRENGTH, ACTIVITY COEFFICIENTS, AND EFFECT ON EQUILIBRIA

If the solution in which the reaction takes place acts as a real solution, which generally occurs with electrolytes in solution, the activity coefficients will result in an "apparent equilibrium constant", K_C', that is different from the thermodynamic value (set by $\Delta G°$). The relationship is given below where the ratio of the activity coefficients, K_γ, is multiplied by K_C' to equal K_{THERM}.

$$K_{THERM} = \frac{a_C{}^c a_D{}^d}{a_A{}^a a_B{}^b} = \frac{\left(\gamma_C[C]\right)^c \left(\gamma_D[D]\right)^d}{\left(\gamma_A[A]\right)^a \left(\gamma_B[B]\right)^b} = \frac{\left(\gamma_C\right)^c \left(\gamma_D\right)^d}{\left(\gamma_A\right)^a \left(\gamma_B\right)^b} \frac{[C]^c[D]^d}{[A]^a[B]^b} = K_\gamma K_C'$$

Although K_{THERM} is fixed by the value of $\Delta G°$ for the reaction, the change in the effective concentration of ions in solution, in terms of ionic strength, does cause a change in the observed or "apparent" equilibrium constant, K_c', measured for the solution. The observed value (K_c') is dependent on the ionic strength.

Ionic strength, I, is a measure of the "ionic atmosphere", involving both the water and the ions in the solution.

Defining Ionic Strength: $I = $ ionic strength $= \sum_{all\,ions} c_i z_i^2$

■ The greater the *ionic strength* of a solution, the higher the charge of the ionic atmosphere.

■ Increasing ionic strength means a net reduction in attractive forces between any ion pair.

■ The concept of bound versus free water molecules affecting activity and then the thermodynamic parameters of the solution discussed in Part 5 also applies to equilibria involving ions.

For neutral species, the activity is independent of ionic strength, but any charged ions will have activities that are dependent on the ionic strength of the solution. The effect of the activity coefficient can be to either lower or raise the apparent equilibrium concentration relative to that in an infinitely dilute solution, where $\gamma_{ion} = 1.0$ for the same temperature. Only by replacing concentration terms with activities by applying the definition: $a_{ion} = \gamma_{ion} C_{ion}$, will the potential effect of the activity coefficient of the ion on K become apparent.

For solutions with ionic strengths of 0.10 M, an estimate of the activity coefficient of the ion can be made using the Debye–Hückel limiting law or the extended Debye–Hückel equation.

Debye–Hückel Limiting Law (DHLL): $\log \gamma_{ion} = \dfrac{-0.509(z_i)^2 \sqrt{I}}{1 + (\alpha / 305)\sqrt{I}}$ where $\alpha = $ hydration radius of ion

■ The law works well for dilute solutions, when $I < 0.10$, and gives a reasonably good approximation at higher ionic strength levels.

■ The activity coefficient, γ_{ion}, is equal to 1.0 in an infinitely dilute solution or ≈ 1.0 in low ionic strength solutions.

■ The hydration radius (α) applies to the ion and the water molecules tightly bound around it. The measured values are given in reference tables.

- If the hydration radius is not known, the approximation: $\log\gamma_{ion} = -0.509(z_i)^2\sqrt{I}$ may be used.
- Values of $\gamma_{ion} < 1.0$ occur in higher ionic strength solutions, and the value depends both on I and the hydration radius of the ion.
- When the magnitude of the charge, z, of the ion increases, its activity coefficient decreases.

   NOTE: *Activity corrections are more important for ions with a charge of ±3 than for ions with a charge of ±1.*

- The smaller the hydrated ion size (α), the more important the activity effects become for the ion.

Many commonly measured equilibrium values are affected by ionic strength. Solution values like pH are affected by ionic strength since:

$$pH_{obs} = -\log(a_{H_3O^+}) = -\log(\gamma_{H_3O^+}[H_3O^+]) = -\log(\gamma_{H_3O^+}) + (-\log[H_3O^+])$$

- Buffer pH values are affected by ionic strength and, since either acid and/or base form may be ions, the pH may change dramatically from the thermodynamically predicted value, given that buffers are normally high ionic strength solutions.
- Solubility equilibria are also greatly affected by the ionic strength resulting from ions, not involved in the equilibrium, influencing the activities of each ion. The solubility of a salt (concentration of dissolved ions) will increase as the activity coefficient of the ions decreases.

EXAMPLE PROBLEMS

6.18   Show how each of the following equilibrium constants would need to be rewritten to define K_{THERM} with the appropriate activity coefficients and K_c.
   A) K_b of the weak base, CH_3NH_2: $CH_3NH_2 + H_2O(l) \Leftrightarrow CH_3NH_3^+ + OH^-$
   B) K_a of the conjugate acid, NH_4^+: $NH_4^+ + H_2O(l) \Leftrightarrow NH_3 + H_3O^+$
   C) K_f of the complex, $Fe(H_2O)_6^{+2}$: $Fe^{+2} + 6\ H_2O(l) \Leftrightarrow Fe(H_2O)_6^{+2}$
   D) K_f of the complex, $Co(Cl)_4^{-2}$: $Co^{+2} + 4\ Cl^- \Leftrightarrow Co(Cl)_4^{-2}$

6.19   A)   Each of the equilibrium reactions described below will be driven to the right (i.e. having a greater tendency to dissociate into ions) if the ionic strength is increased from 0.01 to 0.1 M. Explain why that would be the case, using activity coefficients.

   (a) $CaCO_3(s) \rightleftarrows Ca^{+2}(aq) + CO_3^{-2}(aq)$ (b) $B(aq) + H_2O(l) \rightleftarrows BH^+(aq) + OH^-(aq)$

   B)   The complex $FeSCN^{+2}$ ion is formed with the equilibrium: $Fe^{+3}(aq) + SCN^-(aq) \rightleftarrows [Fe(SCN)]^{+2}$. The observed ratio of complex formed to reactant ions decreases if the ionic strength is increased from 0.01 to 0.1 M. Explain why this would be the case, using activity coefficients.

6.20   Using $BaCrO_4(s)$ as an example [$K_{sp} = 2.11 \times 10^{-10}$], calculate the molar solubility of $BaCrO_4$ in:
   A)   pure water
   B)   0.10 M NaCl, considering activities

6.21   A)   For a weak acid dissociation: $HA + H_2O(l) \Leftrightarrow H_3O^+ + A^-$, prove that, if activities and activity coefficients are included, then $[H^+]_{equil} = \sqrt{K_a[HA]_{initial}/\gamma_{H^+}\gamma_{A^-}}$.
   B)   Calculate the pH in a 0.100 M solution of CH_3CO_2H for its dissociation in:
      (a)   pure water
      (b)   1.0 M KCl

6.22 A) Prove that, when activities are included, the equation for the pH of a buffer becomes the second equation shown below:

$$pH = pK_{a,\text{acid form}} + \log\frac{[\text{base form}]}{[\text{acid form}]}$$

$$\rightarrow pH_{\text{obs}} = pK_{a,\text{acid form}} + \log\left[\frac{\text{mol base}}{\text{mol acid}}\right] + \log\left[\frac{\gamma_{\text{base}}\gamma_{H^+}}{\gamma_{\text{acid}}}\right]$$

B) Suppose you combine 2.00 g $NaH_2PO_4(s)$ [MW = 104] with 9.00 g $Na_2HPO_4(s)$ [MW = 126] in 500 mL of water. What will be the pH in the solution?
(a) Ignoring activities
(b) Including activities.

KEY POINTS – ELECTROCHEMISTRY, EQUILIBRIUM, AND ACTIVITIES

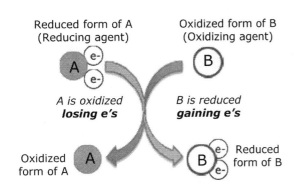

Reduced form of A
(Reducing agent)

A is oxidized
losing e's

Oxidized form of A

Oxidized form of B
(Oxidizing agent)

B is reduced
gaining e's

Reduced form of B

Redox reactions are chemical reactions that can be described as involving a transfer of electrons from one substance (reducing agent) to another (oxidizing agent). One element is reduced (gains electrons) while another is oxidized (loses electrons). The gain of H atoms is often associated with reduction, while adding O atoms to a molecule is oxidation, but we are not limited to those types of reactions only. The ΔG associated with a redox reaction is particularly useful, since electron transport forms an important part of many reactions in chemistry and biology.

Since the ΔG of a chemical reaction represents the maximum non-PV work that can be extracted from the reaction, it gives the magnitude of any electrical work that can be obtained from the reaction. Using Faraday's constant, F, the ΔG for any redox (oxidation-reduction) reaction can be expressed as a voltage, potential difference (ΔE) or electromotive force (emf). The use of various representations of the emf can result in variety of symbols used in the equation for ΔG, as shown below, but the context is usually clear, that a voltage is being used to define the thermodynamics of the electrochemical cell or redox reaction. $\Delta G = -nF\Delta E = -nFE = -nF\Phi$

- Note, however, that the signs of ΔG and ΔE are opposite. When the voltage difference is positive, the ΔG is negative, signaling spontaneous transfer of electrons from anode to cathode.

Since the difference in free energy between the cathode (reduction) and anode (oxidation) half-reactions in the overall reaction defines all thermodynamic terms for the reaction, the convention defining the order of subtraction must be strictly followed, so that no confusion results with respect to the sign of the ΔG.

- The full-reaction voltage, $\Delta E°$, for standard conditions is defined by:

$$\Delta E° = E°_{\text{cathode}} - E°_{\text{anode}} = E°_{\text{red}} - E°_{\text{oxid}}$$

- Therefore standard half-reaction voltages, $E°$, are all defined as reductions when tabled, and the voltage measured against the $2\,H^+ + 2e^- \rightarrow H_2(g)$ half-reaction (set as the "zero" voltage value) under standard conditions, namely 25°C, 1.0 atm (bar) and concentration of dissolved species 1.0 M so that:

$$\text{Oxidized form} + ne^- \rightarrow \text{reduced form} \quad E°\left(\text{tabled value}\right)$$

- The tabled $E°$ for the anode half-reaction is always subtracted, so that the half-reaction becomes reversed, when added to the cathode reaction, and describes oxidation.

A galvanic or voltaic cell has a $\Delta E°$ that is positive, indicating that $\Delta G°$ is negative, and a nearly complete conversion of products to reactants will take place spontaneously. If $\Delta E°$ is negative, the reaction will be non-spontaneous and "weak", having relatively few products at equilibrium.

CHEMICAL EQUILIBRIUM

$$\left[\frac{d\Delta G}{d\xi}\right]_{P,T} = \Delta G = \Delta G° + RT \ln\left[\frac{a_C^c\, a_D^d}{a_A^a\, a_B^b}\right]$$

$$\Delta G = -nF\Delta E (volts)$$

ELECTROCHEMISTRY

$$\Delta E = \Delta E° - \frac{RT}{nF} \ln\left[\frac{a_C^c\, a_D^d}{a_A^a\, a_B^b}\right]$$

- The number of electrons transferred within the overall reaction must be the same for reduction as for oxidation and is represented as "n" in the equation, describing the emf or $\Delta G°$.
- Although the half-reactions will be multiplied by various factors to make sure the same number of electrons is transferred, the $E°$ values MUST NEVER be multiplied by these factors, since the adjustment for stoichiometry was taken care of in the measurement against the $H^+/H_2(g)$ half-cell.

At equilibrium, $\Delta G=0$, so $\Delta E=0$ and no electrical energy can be extracted from the reaction. However, the $\Delta E°$ cannot be zero, since it is a form of $\Delta G°$, but will define the equilibrium constant. In fact, electrochemical measurements are a very precise way to measure equilibrium constants that would be very difficult to determine through other means.

$$\Delta E° = -\frac{\Delta G°}{nF} = \frac{RT}{nF}(\ln K) = \frac{2.303RT}{nF}(\log K)$$

The Nernst equation defines the non-equilibrium cell voltage or emf, ΔE_{cell}, depending on the extent of the reaction. It can be used to determine the direction of spontaneous change for any redox reaction under any conditions, and therefore is an important equation in electrochemistry. If the value of the temperature is assumed to be 298.15 K, then ratio of the constants, RT/F, can be reduced to a numerical value, so that:

$$\text{Nernst Equation: } \Delta E_{cell} = \Delta E° - \frac{RT}{nF}\ln Q = \Delta E° - \frac{0.02569(V)}{n}\ln Q$$

$$= \Delta E° - \frac{0.05916(V)}{n}\log Q$$

In order to produce a flow of electrons through the electrodes, the half-reactions must be separated into cathode and anode compartments by some physical barrier (as discussed in your textbook). The arrangement of the full "cell" is described using a specific **cell notation** as:

$$M_1 \,|\, \text{anode species} \,|\, \text{cathode species} \,|\, M_2$$

where:

- The metal electrodes bracket the notation, starting with the metal used for the anode (M_1) at the beginning and then the metal used for the cathode (M_2) at the end.
- All chemical species in the half-reaction for the ANODE are listed on the left side while species in the CATHODE half-reaction are listed on the right side of the separator, for the two half-reactions.
  - Use commas to separate dissolved species
  - Give the molar concentration or state in parentheses after chemical name
  - Insert a "|" between chemical species separated by a phase boundary
  - Use the symbol "||" for a salt bridge if it is used as a separator.

The proper cell notation should give you enough information to determine the specific half-reactions combined, the value of n in the overall reaction and the $\Delta E°$ for the cell.

When two solutions containing the same ion but at different concentrations, are separated by a semipermeable membrane, a difference in chemical potential, $\Delta\mu$, will exist for that ion, producing a ΔE across the membrane. This situation is described as a **concentration cell**, where the $\Delta\mu$ for the ion can be defined, using standard electrochemistry. The solution with the higher concentration is the cathode, where reduction occurs, while the

Semipermeable Membrane

Anode: $M \rightarrow M^{+nx} + ne^-$	Cathode: $M^{+nx} + ne^- \rightarrow M$
$[M^{+nx}]_{low}$	$[M^{+nx}]_{high}$

$\Delta\mu \rightarrow \Delta E$

Electron "flow" would equalize concentrations

lower concentration solution is the anode, where oxidation occurs. Since the $E°$ terms are identical for both half-reactions, then:

$$\Delta E = E_{cathode} - E_{anode} \text{ leads to } \Delta E = \frac{0.0592}{n} \log \frac{[M^{+x}]_{high}}{[M^{+x}]_{low}}$$

Reduction at the cathode means $[M^{+nx}]$ would be lowered in concentration in that part of the cell, while oxidation would increase $[M^{+nx}]$ in the anode compartment, thereby eventually producing equal concentrations in both.

EXAMPLE PROBLEMS

6.23   Consider the two half-reactions below:

$$ClO_4^- + H_2O(l) + 2e^- \rightarrow ClO_3^- + 2OH^- \qquad E° = 0.36 \text{ V}$$
$$Cd(OH)_2(s) + 2e^- \rightarrow Cd(s) + 2OH^- \qquad E° = -0.81 \text{ V}$$

   A)   In order to produce a spontaneous reaction, which should be reduced, ClO_4^- or $Cd(OH)_2$?
   B)   Write the balanced overall reaction and the form of K for the overall reaction.
   C)   What would be the $\Delta G°$ and K for the spontaneous reaction at 25°C?

6.24   Which of the following redox reactions would produce approximately the same amount of work (±5 kJ) per mole of metal oxidized?
   (a) $Zn(s) + Cu^{+2} \rightarrow Zn^{+2} + Cu(s)$ (b) $Ca(s) + H^+ \rightarrow Ca^{+2} + H_2(g)$
   (c) $Li(s) + H_2O \rightarrow Li(s) + ½ H_2(g) + OH^-$ (d) $2 Ni(s) + O_2(g) + 4 H^+ \rightarrow 2 Ni^{+2} + 2 H_2O$

6.25   Construct an electrochemical cell that would let you measure the K_{sp} of $Hg_2SO_4(s)$ at 25°C and:
   A)   Give the half-reactions for the cathode and anode, and then the $\Delta E°$ for the cell.
   B)   Calculate K_{sp} from the data chosen
   C)   Compare the calculated result for the K_{sp} to the literature value of 6.5×10^{-7}.

6.26   Given the following two half-reactions:

$$Hg(CN)_4^{-2} + 2e^- \rightarrow Hg(l) + 4CN^- \qquad E° = -0.37 \text{ V}$$
$$Hg_2^{+2} + 2e^- \rightarrow 2Hg(l) \qquad E° = 0.86 \text{ V}$$

   A)   How could you add them to produce the appropriate reaction describing the formation of the $Hg(CN)_4^{-2}$ complex from its ions at 25°C?
   B)   Calculate $\Delta G°$ and the value of K for the formation.
   C)   (a)   Give the proper cell notation to describe the cell needed to measure the K of formation.
         (b)   For a typical cell involving $Hg(l)$ as the electrode, a copper wire will be inserted into the $Hg(l)$ in the cell compartment. Explain why this would be necessary.

6.27   For the disproportionation reaction: $2 CuCl(aq) \rightarrow Cu(s) + CuCl_2(aq)$
   A)   What are the two half-reactions involved in the redox reaction and what is the $\Delta E°$ for the overall reaction?
   B)   What is the K for the disproportionation at 25°C?
   C)   If the initial concentration of CuCl is 0.200 M,
         (a)   What would be the cell potential when 50% of the [CuCl] is converted to $CuCl_2$ and Cu(s)?
         (b)   Would the voltage change if the initial concentration were 2.0 M instead of 0.20 M?
   D)   What would happen to the cell potential if the same amount of HCl is added to both the cathode and the anode?

6.28 The permanganate ion is a standard oxidizing agent (where Mn is reduced) that requires very acidic conditions.
A)   What is the half-cell potential of the MnO_4^-, H^+/Mn^{+2} half-reaction at:
(a)   pH = 6.00?
(b)   pH = 2.00?
B)   Which conditions are more favorable for the reduction potential of Mn? Explain your reasoning.

6.29 When $CH_4(g)$ is oxidized completely to $CO_2(g)$ and $H_2O(l)$ by reacting with $O_2(g)$ in fuel cell where the reaction is the same as that for normal combustion of $CH_4(g)$:
A)   Calculate $\Delta E°$ for the fuel cell reaction from $\Delta G°$. A typical fuel cell produces a potential of about 1.00 V. Is this cell in the same range?
B)   What is the half-reaction for $CH_4(g)$ from the combustion equation and value of its $E°$, given that the half-reaction for O_2 could be considered to be: $O_2(g) + 4 H^+ + 4 e- \rightarrow 2 H_2O(l)$.
C)   How much electrical energy can be extracted in the form of electrical work from the oxidation of 30.0 L of $CH_4(g)$ at 20.0 atm and 25°C, with excess O_2?

6.30 If you have the following cell: $Cu| Cu^{+2}|| Fe^{+2}, Fe^{+3}|Pt$ and it contains $Fe(NO_3)_3(aq)$ at an initial concentration of 0.200 M, what would be the molarity of Cu^{+2} when 30% of the Fe^{+3} is converted to Fe^{+2} and the cell voltage is 0.472 V?

6.31 For a concentration cell involving the ion, M^{+x}, where $M^{+x} + xe^- \rightarrow M(s)$
A)   Derive the proper final expression for ΔE of the cell from the Nernst equations for the half- cells.
B)   For the following four concentration cells, which one has the:
(a)   Highest ΔE
(b)   Lowest ΔE
(c)   Approximately the same ΔE

Cell I: $Pt|Cr(NO_3)_3 (0.020\ M)|| Cr(NO_3)_3 (0.300\ M)|Pt$ (membrane permeable to Cr^{+3})
Cell II: $Pt|FeCl_2 (0.200\ M)||FeCl_2 (1.00\ M)|Pt$ (membrane permeable to Fe^{+2})
Cell III: $Pt|SnCl_2 (1.00\ M)||SnCl_2 (2.00\ M)|Pt$ (membrane permeable to Sn^{+2})
Cell IV: $Pt|CoCl_3 (0.232\ M)||CoCl_3 (2.60\ M)|Pt$ (membrane permeable to Co^{+3})

6.32 If the membrane potential of −67 mV is measured for a resting cell, where the membrane is permeable to K^+, what ratio of $[K^+]$ inside versus $[K^+]$ outside would correspond to this voltage at 37°C?

6.33 A voltmeter with the appropriate electrodes reads 100 mV for a standard buffer solution with a pH of 3.0 and the same meter reads 216 mV under the same conditions when placed in a solution of unknown pH. What is the pH of the unknown solution?

6.34 Consider a copper ion concentration cell illustrated on the right. One side contains a 1.00 M $CuNO_3$ solution and the other contains a saturated solution of CuCl(s). If the cell potential is 0.175 V, what is the K_{sp} of CuCl?

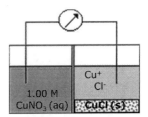

ELECTROCHEMISTRY AND EQUILIBRIUM IN BIOCHEMICAL REDOX REACTIONS

In biochemical systems, oxidation-reduction reactions are essential for energy storage and conversions. Most of these reactions occur in aqueous systems buffered close to pH = 7.0, so both the concentration of water and H^+ are fixed. Consequently:

- A "transformed free energy", $\Delta G'°$, is defined where the additional constraints of pH = 7.0 and fixed concentrations of Mg^{+2} ions and water are added to the usual standard conditions.
- An "apparent equilibrium constant", K', is then defined with the same concentration ratio as K_c, for all species, except the $[H^+]$ term. Since the

"Transformed" Biochemical Nernst Equations:

$$\Delta G' = \Delta G'° + 2.303RT \log Q$$

$$\Delta E' = \Delta E'° - \frac{0.05916}{n} \log Q$$

concentrations of H^+ and Mg^{+2} ions are fixed, the values are included in the value of K, as constants, to produce K'. The concentrations will not appear in the Q or K' expressions.

▪ The "transformed" *standard* electrochemical reduction potentials, E'°, are also measured under the additional constraints of pH = 7.0 and fixed concentrations of Mg^{+2} ions and water, so that they can be related to $\Delta G'^\circ$, using the standard relationships: $\Delta G'^\circ = -nF\Delta E'^\circ$ (or $\Delta G'^\circ = -nFE'^\circ$).

▪ The apparent equilibrium constant, K', can then be calculated from the transformed $\Delta G'^\circ$ value, where pH = 7.0, given that: $\Delta G'^\circ = -RT\ln K' = -2.303RT\log K'$

▪ It will also be true that $K' = K_c \times 10^{7x}$ where "x" is the coefficient of H^+ in the overall reaction.

The value of "x" is negative when H^+ is a reactant, or positive if H^+ is a product. The $\Delta G'^\circ$ may then be calculated from standard thermodynamic values when "x" is known, since:

$$\Delta G'^\circ = \Delta G^\circ - 7x(RT)\ln 10 = \Delta G^\circ - 16.12x(RT)$$

EXAMPLE PROBLEMS

6.35 Consider the half-reactions for $NAD^+/NADH$ and $CO_2/$formate:

$$NAD^+ + 2e^- + H^+ \rightarrow NADH \qquad E'^\circ = -0.320 \text{ V}$$
$$CO_2(g) + 2e^- + H^+ \rightarrow HCO_2^- \qquad E'^\circ = -0.42 \text{ V}$$

A) In order to produce a spontaneous reaction, which should be reduced, NAD^+ or CO_2?

B) What would be the $\Delta G'^\circ$ and K' values for the spontaneous reaction at 25°C?

6.36 Given that the overall reaction is $NADH + O_2(g) + H^+ \rightarrow NAD^+ + H_2O_2(aq)$ for the oxidation of NADH, reduced nicotinamide dinucleotide, by $O_2(g)$ then:

A) What are the half-reactions for the anode and cathode?

B) Write a suitable cell notation for this cell.

C) If the biological reduction E'° for the $NADH/NAD^+$ half-reaction (pH = 7.0) is -0.320 V and that for the O_2/H_2O_2 (pH = 7.0) reduction is $+0.295$ V, what is the:

(a) $\Delta E'^\circ$

(b) $\Delta G'^\circ$

(c) K'

6.37 The terminal respiratory chain involves the redox couples NAD+|NADH and FAD|$FADH_2$.

A) Calculate the $\Delta G'^\circ$ for the following reaction at 298 K: $NADH + FAD + H^+ \rightarrow NAD^+ + FADH_2$

B) Is this $\Delta G'^\circ$ sufficient to synthesize ATP from ADP and inorganic phosphate? Give the reason(s) for your decision.

6.38 The oxidation of lactate to pyruvate by the oxidized form of cytochrome c (Fe^{+3}) is an important biological reaction. The half-cells and relevant E'° values are:

$$pyruvate + 2H^+ + 2e^- \rightarrow lactate \qquad\qquad E'^\circ = -0.185 \text{ V}$$
$$\text{cytochrome c}\left(Fe^{+3}\right) + e^- \rightarrow \text{cytochrome c}\left(Fe^{+2}\right) \qquad E'^\circ = 0.254 \text{ V}$$

A) What's the overall reaction and its $\Delta E'^\circ$?

B) Calculate the value of K' for this reaction.

C) Suppose that, at some point in the reaction, the concentration of the reduced form of cytochrome c (Fe^{+2}) is 1000 times that of the oxidized form, cytochrome c(Fe^{+3}). What's the ratio of [pyruvate]/[lactate] at the same point?

D) Suppose the concentrations in the reaction were: [cytochrome c(Fe^{+2})] = 100 mM, [cytochrome c(Fe^{+3})] = 10 mM, [pyruvate] = 0.200 M and [lactate] = 10 µM; under such conditions, which concentration would increase – pyruvate or lactate? Explain your decision.

6.39 The pH dependence of various biological half-reactions was determined [R.A. Alberty, Biophysical Chemistry (2004) 111, 115–122] so that they can be used to calculate the apparent equilibrium constant for a biological redox reaction. A selection of the data appears in the table on the right for the $NAD^+ \rightarrow NADH$ and xylitol $\rightarrow$ xylulose half-reactions.

Half reaction (298 K)	pH = 5.0	pH = 6.0	pH = 7.0
E°′, NAD+/NADH (V)	−0.2569	−0.2865	−0.316
E°′, Xylitol/xylulose (V)	0.1076	0.1667	0.2259

A)   Calculate the $\Delta E^{\circ\prime}$ and ΔG° values for the spontaneous reaction at a pH of 5, 6, or 7.

B)   How much larger is the K' at pH = 7.0 than K' at pH = 5.0?

C)   If you have the same concentrations of reactants and products, what would be the % increase (or decrease) in the maximum extractable work from the reaction from:
(a) pH 5.0 to pH 6.0?   (b) pH 6.0 to pH 7.0?   c) pH 5.0 to pH 7.0?

6.40 A table of "transformed" free energy of formation values, $\Delta G_f^{\prime\circ}$, for various biochemical substances was developed by Robert A. Alberty [R.A. Alberty, Biophysical Chemistry (2004) 111, 115–122] for different ionic strengths of solutions. Given the values in the table below:

Substance		$\Delta G_f^{\prime\circ}$ (kJ/mol)	
	I = 0	I = 0.10	I = 0.25
cytochrome c(Fe^{+3})	0	−5.51	−7.29
cytochrome c(Fe^{+2})	−24.54	−26.96	−27.75
Pyruvate	−352.4	−351.2	−350.8
Lactate	−316.9	−314.5	−313.7

A)   Calculate the $\Delta G^{\prime\circ}$ from the $\Delta G_f^{\prime}$ values for the reaction given in Problem 6.38 at the three ionic strengths given.

B)   (a)   Calculate the K' values at the different ionic strengths.
(b)   How does the value of K' for $I = 0$ compare to the value calculated in Problem 6.38?

C)   Compare the values of K' by determining how much larger is the K' value at:
(a) $I = 0$ versus $I = 0.10$
(b) $I = 0$ versus $I = 0.25$

D)   Why would the value of K' at different ionic strengths be useful to know for a biochemical reaction in particular?

6.41 An important biochemical redox reaction is the oxidation of ethanol, CH_3CH_2OH, to acetaldehyde, CH_3CHO, by NAD^+. Given the cathode and anode half-reactions below and the $\Delta G_f^{\prime\circ}$ values in the table

Cathode: $NAD^+ + H^+ + 2e^- \rightarrow NADH$
Anode: $CH_3CH_2OH \rightarrow CH_3CHO + 2\,H^+ + 2e^-$

Substance	$\Delta G_f^{\prime\circ}$ (kJ/mol)	
	I = 0	I = 0.25
NAD$^+$	1038.9	1059.1
NADH	1101.5	1120.1
CH$_3$CH$_2$OH	58.0	62.96
CH$_3$CHO	20.83	24.06

A)   Write the balanced overall reaction.

B)   Calculate the $\Delta G^{\prime\circ}$ for the reaction from the $\Delta G_f^{\prime\circ}$ values and then the K' at the two ionic strengths given.

C)   Compare the change in K' with ionic strength to that observed in Problem 6.33. Are the trends the same? Give the reason(s) for your decision.

D)   Using the relationship between K_c and K' given in the key points summary,
(a)   Calculate the value of K_c for this reaction for $I = 0$.
(b)   Is the change in value observed between K' and K_c consistent with what we would expect based on the change in pH and Le Chatelier's principle? Explain the reason(s) for your decision.

6.42 Given the following reduction half-reactions at pH = 7.0: [R.A. Alberty, Biophysical Chemistry 111 (2004), 115–122]:

$$O_2\,(g) + 4e^- \rightarrow 2H_2O\,(l) \qquad E^{\prime\circ} = 0.849 \text{ V}$$

$$Cystine + 2e^- \rightarrow 2\,cysteine \qquad E^{\prime\circ} = -0.363 \text{ V}$$

A)   Write the balanced overall reaction and determine $\Delta E^{\prime\circ}$ and $\Delta G^{\prime\circ}$ for the reaction.

B)   If you prepare a 1.00 mM solution of cysteine in water at 25°C and allow it to sit in a beaker exposed to air ($P_{O2} = 0.20$ atm), what can you expect will happen to the concentration of cysteine in the solution?

KEY POINTS – THERMODYNAMICS OF REDOX REACTIONS: $\Delta S°$ AND $\Delta H°$

Electrochemical measurements can provide the value of thermodynamic parameters such as the $\Delta H°$ and $\Delta S°$ for a redox reaction, which may be difficult to determine through calorimetric measurements. If the temperature dependence of $\Delta E°$ is measured, then both $\Delta H°$ and $\Delta S°$ for the redox reaction can be determined by modifying the van't Hoff equation, $\ln K = -\dfrac{\Delta H_r°}{RT} + \dfrac{\Delta S_r°}{R}$, as shown on the map and in the figure on the right. In addition, applying Kirchhoff's law, for the T dependence of ΔH:

$\Delta H(T_2) = \Delta H(T_1) + \Delta \bar{C}_p \Delta T$, allows the ΔC_p of the reaction to also be determined, as shown on the map and on the right.

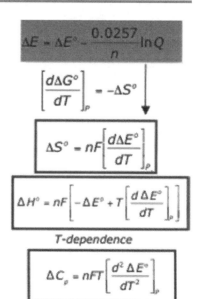

$$\Delta E = \Delta E° - \frac{0.0257}{n} \ln Q$$

$$\left[\frac{d\Delta G°}{dT}\right]_p = -\Delta S°$$

$$\Delta S° = nF\left[\frac{d\Delta E°}{dT}\right]_p$$

$$\Delta H° = nF\left[-\Delta E° + T\left[\frac{d\,\Delta E°}{dT}\right]_p\right]$$

T-dependence

$$\Delta C_p = nFT\left[\frac{d^2\,\Delta E°}{dT^2}\right]_p$$

EXAMPLE PROBLEMS

6.43 Given the cell notation for the voltaic cell: $Zn(s)|ZnCl_2(aq)|$ $AgCl(s)|Ag(s)$:

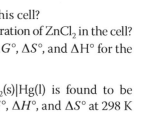

    A)   For the cell, what is the:
        (a)   Balanced overall reaction?
        (b) $\Delta E°$ for the reaction
        (c)   For the schematic of the cell on the right, give the chemical identity of parts indicated by numbers (1)–(4). What is the function of the part(s) labeled (5) on the schematic?
        (d)   Why don't we need a separator or salt bridge for this cell?
    B) ΔE for the cell is measured as 1.015 V; what is the concentration of $ZnCl_2$ in the cell?
    C)   If $(\partial \Delta E/\partial T)_P = -4.02 \times 10^{-4}$ V/K for the cell, what are $\Delta G°$, $\Delta S°$, and $\Delta H°$ for the cell reaction?

6.44 The standard potential of the cell: $Pt|H_2(g)|HCl(aq)|Hg_2Cl_2(s)|Hg(l)$ is found to be +0.2699 V at 293 K and +0.2669 V at 303 K. What are the $\Delta G°$, $\Delta H°$, and $\Delta S°$ at 298 K for this cell?

6.45 Consider the following cell: $Ag(s)|AgCl(s)|NaCl(aq)|Hg_2Cl_2(s)|Hg(l)$ at 25°C and 1.0 atm:

T (K)	$\Delta E°$ (mV)
291	43.0
298	45.4
303	47.1
311	50.1

    A)   Write the anode, cathode, and overall reactions for the cell
    B)   Calculate $\Delta G°$ for the cell
    C)   Given the data measured for the reaction which are shown in the box, calculate $\Delta S°$ and $\Delta H°$ for the cell reaction.
    D)   If the connections remain the same, can this cell change the sign for $\Delta E°$ as T is changed? If yes, are we above or below the changeover temperature at 298 K?

6.46 The $\Delta E°$ values in the table on the right are for an electrochemical cell for which the overall reaction has an $n = 1.0$ and the voltages measured from 280 to 308 K. From the data:

T (K)	$\Delta E°$ (V)
280	0.23302
284	0.23085
288	0.22857
292	0.22619
296	0.22371
300	0.22112
304	0.21843
308	0.21564

    A)   Calculate the $\Delta G°$ and $\ln K$ for each $\Delta E°$ value
    B)   Plot $\Delta E°$ versus T(K) to determine $S°$ for the reaction. Is the resulting plot linear, indicating that $\Delta S°$ is constant over the temperature range? If so, calculate the value of $\Delta S°$ for the reaction.

C)   Plot ln K versus $1/T(K)$ to determine $\Delta H°$ for the reaction. Is the resulting plot linear, indicating that $\Delta H°$ is constant over the temperature range? If so, calculate the value of $\Delta H°$ for the reaction.

D)   The $\Delta G°$ values appear to vary somewhat. Calculate $\Delta H°$ from the $\Delta E°$ values, using the definition given on the map and in key points sections. How do the values obtained compare to the $\Delta H°$ determined in C)? Do they agree?

6.47   The thermodynamics of the dissociation of HBr in solution, where the solvent is a mixture of ethanol and water, has been studied using an electrochemical cell [M.M. Elsemongy, A.S. Fouda, Electrochimica Acta (1981) 26, 255–260] Some of the data collected are given in the table below. The cell notation for the cell is: $Pt|H_2(g)$ $(P = 1.0\ atm)|HBr\ (ethanol + water)|AgBr(s)|Ag(s)$

A)   Write the balanced overall reaction for the cell and the form of K for the overall reaction, given $P\ O_2(g) = 1.0$ atm and $n = 1.0$.

B)   (a)   Calculate the $\Delta G°$ values for each temperature and solution mixture (organizing the results as in the data table).

(b)   What is the effect of increasing temperature on $\Delta G°$ for the reaction in pure water? What does this tell you about the sign of ΔH for the reaction?

(c)   What is the effect of changing the solvent to the mixture on the magnitude and sign of $\Delta G°$?

C)   (a)   Calculate K for the reaction for each value of $\Delta G°$.

(b)   How is the dissociation constant affected by the change in solvent? How much greater is K at the lowest T than at the highest T? Is the trend consistent through each mixture?

D)   Make a plot of $\Delta E°$ versus $T(K)$ for the three solvent mixtures.

(a)   Is the plot linear indicating constant $\Delta S°$ over the temperature range?

(b)   If not linear, estimate the average values within each 10-K temperature region, employing the method used in Problem 6.44. Are there differences in the $\Delta S°$ among the three solvents?

T (°C)		$\Delta E°$ measured, mV	
	Pure H_2O	20% EtOH	50% EtOH
15	75.67	66.29	55.56
25	71.05	62.78	51.54
35	65.85	58.06	45.82
45	60.03	52.23	38.33
55	53.65	47.15	29.18

MEAN IONIC ACTIVITY AND REDOX REACTIONS

Electrochemical measurements can provide the precision needed to determine activity coefficients for ionic species in a redox reaction. However, it may not be possible to separate the activity coefficient of the cation from that of the anion in the cell emf. Consequently, several new terms are defined to express the average of "mean activity" properties for both ions. The symbol ν_+ represents the subscript for the cation in the ionic compound, the symbol ν_- is the subscript for the anion in the ionic compound, and $m\pm$ is the "mean" molality of the electrolyte.

■   Mean activity coefficient $\gamma_\pm = (\gamma_+^{\nu+}\gamma_-^{\nu-})^{1/\nu}$ where $\nu = \nu_+ + \nu_-$

■   Mean ionic molality is defined as: $m_\pm = (m_+^{\nu+}m_-^{\nu-})^{1/\nu}$

■   Mean ionic activity is defined in place of concentration as:
    $a_{electrolyte} = a_+^{\nu+}a_-^{\nu-} = a_\pm^\nu$ or $\gamma_\pm m_\pm = a_\pm$

For 1:1 electrolytes:

$$\gamma_\pm = (\gamma_+\gamma_-)^{1/2} = \sqrt{\gamma_+\gamma_-}$$

2:1 electrolytes:

$$\gamma_\pm = (\gamma_+^2\gamma_-^1)^{1/3} = \sqrt[3]{\gamma_+^2\gamma_-^1}$$

Since the hydration radius cannot be applied to the "mean" coefficient, to estimate the $\log \gamma_\pm$ from the ionic strength, a simplified Debye–Hückel limiting law, as shown below, is used or the Davies equation.

Debye–Hückel: $\log \gamma_\pm = \dfrac{-0.509|z_+z_-|\sqrt{I}}{1+\sqrt{I}}$

Davies Equation: $\log \gamma_\pm = 0.509|z_+z_-|\left[\dfrac{-\sqrt{I}}{1+\sqrt{I}} + 0.300I\right]$

Generally, the Debye–Hückel law applies to solutions that have an $I < 0.10$, whereas the Davies equation applies to more concentrated solutions, with $I < 0.50$ as its upper limit.

EXAMPLE PROBLEMS

6.48 Ephedrine sulfate is an ionic compound with two molecules of ephedrine per one sulfate ion, as shown in the structure on the right.

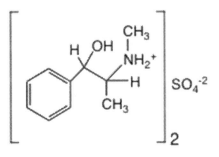

    A) Calculate the mean ionic activity coefficient and mean ionic activity of a 0.0050 m solution of ephedrine sulfate in water.

    B) Calculate the mean ionic activity coefficient and mean ionic activity of a 0.0050 m ephedrine sulfate solution, combined with 0.010 m KCl in water.

    C) Would the results be significantly different in B if you had used the Davies equation? Explain your decision.

6.49 What is the mean ionic activity coefficient and mean ionic activity of 0.20 m K_2SO_4, using the:

    A) Debye–Hückel equation?

    B) Davies equation

6.50 The CRC Handbook of Chemistry and Physics gives the measured mean activity coefficients for various 1:1 and 2:1 electrolytes at various molalities. For the salts $BaBr_2$ and CsF:

    A) Compare the values given in the table (from 88th edition or newer) to those that would be calculated from the:

m (mol/kg)	0.001	0.005	0.010	0.100	0.500	1.000
$\gamma\pm$ $BaBr_2$	0.881	0.785	0.727	0.517	0.435	0.470
$\gamma\pm$ CsF	0.965	0.929	0.905	0.792	0.721	0.726

        (a) Debye–Hückel equation

        (b) Davies equation

    B) Which equation produces the best fit and up to what concentration?

6.51 For aqueous solutions of $NaBrO_3$ and $CuBr_2$:

    A) If a 0.500 m solution of $NaBrO_3$ has a mean ionic activity coefficient of 0.605 at 25°C, what is its mean molality, $m\pm$, mean ionic activity, $a\pm$, and activity for the compound, where $a = a_{\pm}^{v}$?

    B) If a 0.200 m solution of $CuBr_2$ has a mean ionic activity coefficient of 0.523 at 25°C, what is its mean molality, $m\pm$, mean ionic activity, $a\pm$, and activity for the compound, where $a = a_{\pm}^{v}$?

6.52 The solubility limit for $BaCl_2$ in water is 370.43 g per kilogram water at 25°C.

    A) What is the molality of the $BaCl_2$ in the saturated solution?

    B) What is the ionic strength of the solution?

    C) Given that the K_{sp} for $BaCl_2$ is 176.94, what is the mean ionic activity coefficient of $BaCl_2$ in the saturated solution?

6.53 For the cell: $Pt|H_2(g)$ ($P = 1.0$ bar), HCl ($m = 1.00$ m)$|AgCl(s)|Ag$, the measured voltage was 0.214 V. What is the mean ionic activity coefficient for HCl in the solution?

6.54 For the cell: $Zn|ZnCl_2$ ($m = 5.00 \times 10^{-3}$ m)$|Hg_2Cl_2(s)|Hg$, the measured voltage was 1.2272 V. Given that the $\Delta E°$ for the Zn^{+2}/Zn half-cell is 0.7628 V and that for $Hg_2Cl_2(s)|Hg$ is 0.2676 V, what is the mean ionic activity coefficient, $a\pm$, for $ZnCl_2$ in the solution?

6.55 Given the following half-reactions:

$$Cd^{+2} + 2e^- \rightarrow Cd(Hg) \qquad E° = -0.3521 \text{ V}$$
$$AgCl(s) + e^- \rightarrow Ag(s) + Cl^- \qquad E° = 0.2223 \text{ V}$$

    A) Write the overall cell reaction for the spontaneous reaction and its cell notation.

    B) If the concentration of the $CdCl_2$ placed in the cell is 0.010 m, the mean activity coefficient was found to be equal to 0.679. What was the measured cell voltage?

Final Answers

Answers to Example Problems

FINAL ANSWERS AND HINTS

PART 1: GASES AND GAS LAWS

1.1 A) (a) $P_{ideal} = 19.4$ atm   (b) $P_{VdW} = 18.7$ atm   (c) $P_{virial} = 18.7$ atm
- Note the van der Waals and virial equations produce values close to the ideal gas value.

   B) (a) $P_{ideal} = 194$ atm (b) $P_{VdW} = 152$ atm (c) $P_{virial} = 128$ atm
- Note the values are very different under these conditions.

1.2 A) See proof in the Full Solutions at https://www.crcpress.com/Thermodynamics-Problem-Solving-in-Physical-Chemistry-Study-Guide-and-Map/Murphy/p/book/9780367231163.

   B) In 2.0 L, $Z = 0.966$ and in 200 mL, $Z = 0.657$

   C) The attractive forces have increased, given that Z value is much lower than 1.0.

1.3 A) (a) Derivation shown in Full Solutions   (b) Derivation shown in Full Solutions

   B) (a) Derivation shown in Full Solutions

   (b) The observed density of the gas would be greater than if it were an ideal gas.

   (c) The observed density of the gas would be less than if it were an ideal gas.

1.4 A) (a) $d_{ideal} = 78.5$ g/L   (b) $d_{obs} = 89.1$ g/L

1.5 A) 167 atm   B) 295 atm

1.6 A) $Z(I) = 0.934$, $Z(II) = 0.814$

   B) Both values are less than one, so attractive forces dominate. Increasing P increases the attractive forces in NH_3.

   C) $T_{Boyle} = 1368$ K

1.7 A) $R = 0.08206$ L atm/K mol

   B) Calculated values: T_B, $CH_4 = 650$ K; T_B, $N_2 = 438.5$ K; T_B, $H_2 = 113$ K; T_B, Ar $= 516$ K
   Only the value for H_2 comes out close to the tabled value. All other values are overestimates.

1.8 HINT: Have to determine MW from ideal gas law. MW $= 46.0$ g/mol

1.9 A) HINT: Have to determine MW from ideal gas law.
   The value $n = 8$ B) Molecular species is S_8

1.10 A) $Cl_2(g)$ B) 37.2% by mass Ar(g)

1.11 A) Reduced variables have no units   B) Each term has no units   C) 0.0266
   D) $Z \approx 1.0$, so acting ideally   E) 1.64 atm

1.12 A) $V_c = 0.0678$ L mol^{-1} $P_c = 54.5$ atm $T_c = 120$ K   B) $Z_c = 0.375$

1.13 A) $b = 0.0460$ L/mol   B) $Z = 0.661$

1.14 A) No, since $P_{ideal} = 24.45$ atm   B) No, since $P_{VdW} = 21.5$ atm   C) (a) –29.3°C (b) 7.3°C

1.15 A) 3.37 atm   B) No, the original pressure cannot be determined from the given information if a van der Waals gas. See discussion in Full Solutions.

1.16 A) See derivation shown in Full Solutions.
   B) (a) 0.124 L/mol   (b) 0.110 L/mol
   C) A real gas since the van der Waals value is very different from the ideal gas value.

1.17 A)   Quadratic equation results. See derivation shown in Full Solutions.
    B)   Quadratic equation results. See derivation shown in Full Solutions.
    C)   Ideal gas density = 2.88 g/L, van der Waals density = 2.91 g/L
        Virial equation density = 2.94 g/L

1.18 A)   21.2 atm   B)   19.6 atm

1.19 Ratio mol O_2/mol CO @ 50 ppm = 4.58, @ 800 ppm = 0.288, @ 3200 ppm = 0.0712

1.20 A)   4.74 g C_4H_{10} 62.1 g Ar   B) $d = 1.67$ g/L

1.21 A)   Mol %: $H_2 = 80.0\%$ $O_2 = 20.0\%$

1.22 A)   (a) $\chi CO_2 = 0.927$ $\chi_{Ar} = 0.0725$
        (b) $P\,CO_2 = 49.75$ atm $P\,Ar = 3.89$ atm
        (c) $P_{total} = 53.64$ atm
    B) $P\,CO_2 = 40.8$ atm $P\,Ar = 3.87$ atm $P_{total} = 44.67$ atm
    C)   The far fewer moles of Ar gas per L means that the "a" and "b" corrections are very small and the gas acts ideally, since the molecules are widely dispersed in the container.

1.23 A)   4.74 g C_4H_{10}, 61.1 g Ar   B)   1.67 g/L

1.24 A) $P_{total} = 218$ torr   B) $V_{mix} = 3.05$ L

1.25 A) 8.74×10^{-10} atm   B) 8.74×10^{-8} mol %

1.26 A) $\chi HC_1 = 0.0257$ $\chi H_2 = 0.0945$ $\chi Ne = 0.879$
    B) $P\,HCl = 3.55$ kPa $P\,H_2 = 13.0$ kPa $P\,Ne = 121$ kPa
    C)   51.57 g

1.27 A) $P\,CO_2 = 0.710$ atm $P\,H_2 = 0.212$ atm $P\,Ar = 0.511$ atm
    B) $\chi CO_2 = 0.495$ $\chi H_2 = 0.148$ $\chi Ar = 0.357$

1.28 A) $\chi_{N_2(g)} = 0.808$ $P_{N_2} = 0.787$ atm $= 598$ torr

    B) $\chi_{O_2(g)} = 0.190$ $P_{O_2} = 0.185$ atm $= 140.6$ torr

1.29 $V = 4.865 \times 10^4$ L

1.30 A) 220 L/da CO_2 B)   110 L/da O_2 C)   249 kg Na_2O_2

PART 2: FIRST LAW OF THERMODYNAMICS – WORK (PV), HEAT, ΔU AND ΔH

2.1   HINTS: A)   Have to consider whether the volume changes during the step to determine work.
        ■ If either P or V is constant, then T must change since n is constant.
        ■ $\Delta U = 0$ only true when $w = -q$
    B) Derive appropriate equations; be careful with the units.

  A)   (1)   Work done in steps 2, 4 when V changes
      (2) T changes in all steps
      (3)   None, since T changes in each step, none is isothermal
  B) STEP 2: $w = -16.2$ kJ Step 4: $w = -14.6$ kJ
      Need to determine T_{final} to calculate q in each step.

	STEP 1	STEP 2	STEP 3	STEP 4	STEP 5
T_{final}	203.1 K	609.3 K	356.6 K	731.2 K	243.7 K
q	−585 J	+8445 J	−3041 J	+7601 J	−6082 J
THEN:	(1) **w** total	−30.8 kJ	(2) **q** total	6.34 kJ	(3) $\Delta U = -24.5$ kJ

  C) At the end of step 5 = 243.7 K (−29.5°C)

2.2   HINTS: ▪ Know that T and n are constant, so P changes with V.
   A)   "Minimum" will mean it needs irreversible work. Derive appropriate equation, and watch units.
   B)   "Maximum" will mean need reversible work. Must integrate appropriate equations and consult map.
   A) $w = -1.69$ kJ, $q = +1.69$ kJ, $\Delta U = 0$ STEP 2: $w = 0$, $q = 748$ J, $\Delta U = 748$ J

2.3   HINTS: ▪ Irreversible work (against constant P). Use derived equation to solve for P.
   A) $P = 5.00$ atm
   B)   Since isothermal, $q_{irrev} = -w_{irrev}$ to keep T the same and $\Delta U = 0$, $q = +5065.8$ J $= 5.065$ kJ
        $\Delta H = 0$ since $\Delta T = 0$ since n is constant and given $\Delta H - \Delta U = \Delta(nRT) = 0$.
   C) $T_2 = 914$ K $= T_1$

2.4   A) $\Delta U = 4329$ J   B) $\Delta H = 5992$ J   C) $\Delta H - \Delta U = nR(\Delta T) = 1663$ J

2.5 $w = +20.3$ kJ $q = -20.3$ kJ, $\Delta U = 0$ and $\Delta H = 0$

2.6   HINTS: ▪ Integrate function of P to derive reversible work then substitute P ratio for V ratio.
        ▪ For (B) must calculate V_1 and V_2 to get ΔV if you didn't calculate it in (A).
   A)   Greatest work lost by system = Isothermal reversible process, $w_{rev} = -28.6$ kJ
   B)   Least amount of work lost = Isothermal irreversible process, $w_{irrev} = -15.1$ kJ

2.7   HINTS: ▪ Constant P compression, so irreversible work. $w_{irrev} = +288$ J
        ▪ For (B), If no. of moles is the same, then starting P, T sets V_1
        ▪ Use magnitude of work to calculate ΔV given P, then V_2. $V_2 = 0.394$ L

2.8   A) $w_{ideal\ gas} = -4987$ J
   B)   ▪ Integrate to derive reversible work for virial equation, using first two terms only.
        $W_{real\ gas} = -4995$ J, so difference is very small. Only significant when B is very large.

2.9   A) 2.22 atm   B)   −3.63 kJ   C)   −5.72 kJ
   D) See derivation in Full Solutions, work calculated from P ratio $= -5.72$ kJ.

2.10 A) 29.93 J/mol K   B)   21.62 J/mol K

2.11 A)   (a)   Yes, Δn not equal to zero, expansion occurs if $H_2O(g)$ considered product, contraction occurs if $H_2O(l)$ formed instead.

        (b)   Up, if $H_2O(g)$ considered product, down if contraction occurs with $H_2O(l)$ as product.
        (c) P can remain constant since moles and volume change.
   B)   (a)   It will change ΔU since volume cannot change when Δn occurs.
        (b) $\Delta U =$ heat flow in Flask B $= q_V$ and cannot equal the heat flow into the system for Flask A which is equal to $\Delta H = q_P$. Since $q_P > q_V$, the heat flow in Flask A is greater than in Flask B.
   C)   They would be the same since ΔU is a state function, independent of pathway.

2.12 A) $w_{adiab} = -1908$ J   B) $P_{final} = 4.59$ bar

2.13 A)   39.7°C   B)   197°C   C)   Work for A: 405 J   Work for B: 6951 J

2.14 A) $V_2 = 31.4$ L   B) $w_{adiab} = -1.91$ kJ   C) $w_{rev} = -2.30$ kJ

2.15 A) $C_v = 62.9$ J/K mol   B) $\gamma = 1.132$

2.16 A)   See derivation in Full Solutions

   B) $\bar{C}_p = 82.48 \dfrac{J}{mol\ K}$ compared to: $\dfrac{\bar{C}_p}{C_v} = 1.132 \Rightarrow \bar{C}_p = 71.20 \dfrac{J}{mol\ K}$

   C) $Z = 0.9677$

2.17 A)   Assume C_p air $= C_p$ $N_2(g)$. Derive appropriate expression, consult map. $T_2 = 452$ K $= 179°C$

      B)   Increase in ΔU becomes increase in kinetic energy of molecules since tire rubber acts as thermal barrier.

2.18 A)   0   B) 4.125 kJ   C) 4.125 kJ   D) 11.8 L   E) 5.23 atm   F) 5.37 kJ

2.19 A) $\Delta U_{comb} = -30.6$ kJ/g

      B)   ■ Need to calculate ΔH for combustion reaction from ΔU and determine Δn from balanced reaction. $\Delta H_{comb} = -1778$ kJ/mol

      C)   Need to know both molar mass to calculate moles and the chemical formula to determine Δn for the combustion.

2.20 A)   ■ Need to calculate ΔU for combustion reaction from ΔH.
             $C_{cal+contents} = 24.13$ kJ/K

      B)   ■ Need to calculate new ΔU for combustion reaction from new ΔH.
             $C_{cal+contents}$ becomes 25.4 kJ/K, so a small increase.

2.21 A)   ■ Reaction loses heat, so q lost is defined by ΔH.
         ■ Water gains heat, but only changes T, so q_{gain} defined by heat capacity of water and ΔT
             $T_2 = 92.6°C$

      B)   ■ Reaction gains heat so q lost defined by ΔH and moles reacted.
         ■ Water loses heat, so q_{lost} defined by heat capacity of water and ΔT.
             mass NH_4NO_3 needed $= 28.6$ g which is well below solubility limit, so all dissolves.

      C)   Since there are no gases in the reactions occurring, $\Delta n = 0$ and $\Delta U = \Delta H$. (Also, although sealed, plastic sides are flexible so volume could change to keep pressure constant.)

2.22 Mixture: 81.8% CH_4, 18.2% C_2H_6

2.23 A) $C_3H_8(g) + 5O_2(g) \rightarrow 3CO_2(g) + 4H_2O(l)$ $\Delta H_r° = -2210.2$ kJ   B) 88.9%

2.24 A)   6.28 kg
      B)   4.29 kg
      C)   The heat absorbed by warming the melted ice from 0°C to 37°C decreases the amount of ice that would consume 300 calories by about one-third, so it should not be ignored.

2.25 1.0 mol $Ca(OH)_2$, 0.924 mol water vapor, 3.526 mol liquid water, all at 100°C

2.26 A) $\Delta H°_f$ $C_{16}H_{34} = -455$ kJ/mol
      B) $\Delta H°_{comb}$ $C_{16}H_{34} = -1.0728 \times 10^4$ kJ/mol
      C)   At 11.3 kcal/g it has a higher caloric value than the average fats (lipid) at 9.0 kcal/g

2.27 A) $C_{16}H_{34}(l) + 24.5O_2(g) \rightarrow 16CO_2(g) + 17H_2O(l)$ B) -591.5 kJ/mol

2.28 A) See Full Solutions   B) -36.98 kJ

2.29 $\Delta H°_{fermentation} = -70.4$ kJ/mol $C_6H_{12}O_6$
      To produce overall reaction $C_6H_{12}O_6(s) \rightarrow 2CH_3CH_2OH(l) + 2CO_2(g)$ will need to keep $C_6H_{12}O_6(s)$ on reactant side, but $C_2H_5OH(l)$ must appear on the product side, so the second reaction must be reversed before it is added to the first. Also, to get correct coefficient for $C_2H_5OH(l)$, the second reaction must be multiplied by 2.

2.30 A) $\Delta H_r = 499$ kJ for $H_2O(g) \rightarrow H(g) + OH(g)$
         ■ Focus on OH, H and H_2O as the unique substances, which will tell you how to use reactions 1–3.
         ■ Will need to get $H_2O(g)$ on the reactant side, so Reaction 2 must be reversed, and its ΔH multiplied by −1.

- OH is already on correct side and has the needed coefficient, so that reaction can be added as is.
- H is already on correct side but does not have the needed coefficient, so that reaction will be added after multiplying it (and its ΔH) by ½.

B)   Reactions 1 and 2 are formation reactions, while reactions 3 and 4 are dissociation reactions.

2.31 A) $2B(s) + 3H_2(g) \rightarrow B_2H_6(g)$ $\Delta H = \Delta H^\circ_{f,B_2H_6(g)} = -97.4 \text{ kJ}$
- We need $B_2H_6(g)$ on the product side so the first combustion reaction must be reversed and the ΔH multiplied by –1.
- In the second reaction, B is already on the correct side and has the needed coefficient, so that reaction can be added as is.
- Add the formation reaction for $H_2O(l)$ (and its ΔH) but the reaction will be multiplied by 3 before adding.

B)   Using heats of combustion and Hess's law, gives $\Delta H^\circ_{f,B_2H_6(g)} = -97.4 \text{ kJ}$

2.32 Use reactions I, II, III, IV and VI $\Delta H = \Delta H_I + \Delta H_{II} + \Delta H_{III} - 2\Delta H_{IV} - \Delta H_{VI} = -283.5 \text{kJ}$

2.33 A) $\Delta H_r^\circ = -211 \text{ kJ}$ B) See Full Solutions, B.E, O-O $= 143.5 \text{ kJ}$ C) ΔH_r° with enzyme $= -188 \text{ kJ}$

D)   Since bond energies only apply to gaseous molecules forming gaseous atoms, the inclusion of H_2O_2 as an aqueous species and liquid water in the enzyme reaction would make the estimation of the bond energy much more uncertain.

2.34   - NOTE: Since bond energies are for gaseous species only, must condense the gas to the solid to complete pathway, so will $\Delta H_{sublimation}$ for $XeF_6(s)$. Answer: B.E. Xe-F $= 130 \text{ kJ/mol}$

2.35 A) $q_1 = $ B.E. for $N \equiv N$ (or $\Delta H_f^\circ N(g)$) $q_2 = $ B.E. for H-H (or $\Delta H_f^\circ H(g)$) $q_3 = $ B.E. for N-H

B)   Calculated $\Delta H_f^\circ = -52.5 \text{ kJ}$ which is higher than tabled value
C)   H-bonding may be occurring as an attractive force between the gas molecules, which could affect the ΔH value.

2.36 A) $q_1 = \Delta H_{sublimation} C(s)$ $q_2 = $ B.E. for H-H (or $\Delta H_f^\circ H(g)$)
$q_3 = $ B.E. for C-C and B.E. for C-H $q_4 = \Delta H_{vaporization} C_8H_{18}$
Calculated $\Delta H_f = q_1 + q_2 + q_3 + q_4 = -283 \text{ kJ}$
B) ΔH_f° from Hess's Law and $\Delta H_{comb} = -250 \text{ kJ}$
C)   If B.E. for C–C and C=C used, the $\Delta H_f^\circ = q_1 + q_2 + q_3 + q_4 = +1671 \text{ kJ}$, so grossly in error. The comparison to the value in (B), which will be much more accurate, says you made the wrong assumption.
D)   The calculated total change in bond energy for the reaction will not match the true ΔH for the reaction, but we may get a reasonable approximation of the magnitude of the bond energy, so it is only an estimate.

2.37 A)   See Full Solutions for cycle
B)   Calculated value $= 826.8 \text{ kJ}$, so it is close to tabled values.

2.38 A)   Units: $a = $ J/mol K $b = $ J/mol K^2 $c = $ J/mol K^3
B)   123.7 J/mol K   C)   120.5 J/mol K   D)   No terms can be neglected

2.39 A)   –235.4 kJ   B) –234.1 kJ So both produce same result, even though forms are different.

2.40 A)   250.7 kJ   B)   (a) See Full Solutions   (b)   257.7 kJ

C)   Result $= 253.6 \text{ kJ}$, so no significant change when T dependence of C_p taken into account.

2.41 A)   –37.7 kJ   B)   –18.8 kJ

C)   Assuming C_p is constant for C(s) is not a good approximation, $T_2 \gg T_1$ and ΔC_p is large.

2.42 A)   (a)   Assume ΔH_f values of solids can be used for alanine and aspartic acid instead of the dissolved aqueous solution values to calculate the ΔH_r° although the conversion likely takes place in solution phase. This assumes any shift in the ΔH value for alanine and aspartic acid, due to changes in interactions between solute and solvent, are similar and would cancel out.

(b)   Another would be that the C_p values stay constant from 25 to 50°C is necessary, since data for the temperature dependence of the heat capacity alanine and aspartic acid would be difficult to find.

(c)   You would need the $\Delta H_{f\,values}$ of alanine(s), aspartic acid(s) and $CO_2(g)$ and their respective $C_{p,298}$ values. You would also need to use Hess's law and the T dependence of ΔH (Kirchhoff's Law)

B)   16.2 kJ

2.43 A)   –113.1 kJ   B)   113.1 kJ   C)   No, not for this reaction.

2.44 A)   –92.22 kJ   B) $\Delta H(T_2) = -111.9$ kJ, which is a significant change.

2.45 A)   Pathway using cycle shown in Full Solutions or use Kirchhoff's Law $\Delta H_r = 36.77$ kJ
B)   38.2 kJ, so very close result. Can describe as a chemical conversion.

2.46 A)   Alkane n value determined from ΔC_p by using Kirchhoff's law and the two known ΔH values.
Molecular formula $= C_5H_{12}$

PART 3: SECOND AND THIRD LAWS – ENTROPY

3.1 A) ΔS_{vap} CCl_4 <u>85.8</u> H_2S <u>87.9</u> CH_4 <u>73.4</u> H_2O <u>109</u> J/mol K
B)   The values for CCl_4 and H_2S are very similar, indicating the increase in disorder at the liquid–gas conversion is similar while CH_4 is similar but lower. Both CCl_4 and CH_4 have LDF (London Dispersion Forces) interactions only in the liquid, but MW of CCl_4 is much higher, which would increase the LDF. H_2S has more attractive dipolar forces, but the higher MW of CCl_4 has again increased its LDF.
C)   The value for H_2O liquid to gas conversion is much higher, indicating the higher degree of organization of the H-bonding forces in the liquid versus LDF or dipolar forces.

3.2 A) ΔS_{fus} $CH_3C_6H_5$ (37.3 J/mol K) $> \Delta S_{fus}$ C_6H_6 (37.3 J/mol K)
B) ΔS_{vap} C_6H_6 (87.2 J/mol K) $> \Delta S_{vap}$ $CH_3C_6H_5$ (86.7 J/mol K)

3.3 A) $\Delta S_{surr} = 47.2$ kJ/K
B)   Although heat flows in to keep T, the system isothermal, there is no ΔT in the surroundings.

3.4 A) ΔS (constant V) = 6.375 J/mol K
B) ΔS (constant P) = 10.62 J/mol K
C)   Since $q_V \neq q_P$ the ΔS values will not be equal.

3.5 A)   45.5 J/mol K   B)   57.2 J/mol K   C)   The value in (B) is more accurate

3.6 A)   See the Full Solutions for the diagram and equations for each segment below.

Section	Heat solid	Melt solid @ T_{mp}	Heat liquid	Boil liquid @ T_{bp}	Heat vapor
T	30.0 → 118.8°C	118.8°C	118.8 → 185.4°C	185.4°C	185.4 → 200°C
ΔS	14.6 J/mol K	40.3 J/mol K	13.6 J/mol K	91.2 J/mol K	6.7 J/mol K

B) $\Delta S_{total} = 166$ J/mol K   C)   Boiling the liquid is the largest ΔS with 55.3% of the total.

3.7 A) ΔS C(s), 25 → 600°C = 24.3 J/mol K

B) ΔS Fe(s), $25 \to 600°C = 33.0$ J/mol K

C)   Using C_p function, ΔS Fe(s), $25 \to 600°C = 27.0$ J/mol K
   ■ Since significantly lower, integration necessary.

3.8   A) ΔS_{diss} Br   <u>104.5</u>   Cl   <u>107</u>   I   <u>101</u>   H   <u>98.7</u>   J/K

B)   All four values are similar, indicating the most important factor in the ΔS is not chemical identity but that one molecule breaks into 2 atoms.

3.9   ■ Will need to calculate no. of moles in sample; assume ideal gas.
   ■ Breakdown process into 2 reversible steps:
      (1) Volume change at constant T and then
      (2) Let T change at constant final volume yields $\Delta S_{total} = \Delta S_{(1)} + \Delta S_{(2)} = 0.118 + 0.0581 = \underline{0.176}$ J/K

3.10   A)   (a) $\Delta S_{gas} = 3.81$ J/K   (b) $w = -1140$ J
       (c) $\Delta U = 0$, $\Delta S_{surr} = -3.81$ J/K   (d) $\Delta S_{total} = 0$

B)   (a) $\Delta S_{gas} = 3.81$ J/K   (b) $w = 0$
       (c) $\Delta S_{surr} = 0$ (d) $\Delta S_{total} = 3.81$ J/K

3.11   ■ Breakdown process into 2 reversible steps:

A) $\Delta S_{per\ breath} = 0.0350$ J/K   B) $\Delta S_{per\ day} = 1.01$ kJ/K   C) T change

3.12   A)   14.35 J/K   B)   Changing volume doesn't affect ΔS, since mol fraction stays the same.

3.13   ■ Will need to calculate no. of moles in sample; use volume ratio to define molar ratio.

A)   11.42 J/K

B)   Total ΔS is lower because mol fraction of $C_2H_6(g)$ is larger, meaning less choices for possible distribution of particles in the mixture.

3.14   A)   8.91 atm

B)   Reasonable assumption since low pressures in a large volume

C)   66.17 J/K

D)   Would affect mol fraction, if mol ratio not calculated from masses, since V or P ratio may change.

3.15   A) $\Delta S_{mix} = 215$ J/K   B) $\Delta S_{mix} = 11.52$ J/K ΔS_{mix} (A) $= 6.46$ J/K

3.16   A) $\Delta S_{overall} = 32.73$ J/K

3.17   A)   Diagram shown in Full Solutions.   B) 23.4 J/K   C) $\Delta \bar{C}_p = \left[\bar{C}_{pH_2O(s)} - \bar{C}_{pH_2O(l)} \right]$

3.18   A) $\Delta H_{den}(T_2) = \Delta H_{den}(T_m) + \Delta \bar{C}_p(T_2 - T_m)$ $\Delta S_{den,310} = \Delta S_{den,T_m} + \Delta \bar{C}_p \ln(T_2/T_m)$

B)   (a) 1.880 kJ/K mol   (b) 1.109 kJ/mol K

3.19   A)   1.20 kJ/mol K   B) 9.84 kJ/mol K

C) ΔS_{fus} $H_2O = 53.0$ J/K ΔS_{fus} $C_6H_6 = 53.0$ J/K ΔS_{fus} $CCl_4 = 9.99$ J/K ΔS_{fus} $CH_4 = 10.4$ J/K

3.20   A)   0.477 kJ/K mol   B)   The values are strongly affected.

C)   The entropy values for protein transition quoted in kJ/mol, while ΔS for transition values for molecular substances are in J/mol, reflecting hundreds of atoms changing position.

3.21   A) -974.6 J/K   B)   24.9 J/K

3.22   A)   Expect positive and large ΔS_r since forming 3 mol gas from solid and liquid.

B)   424 J/K

3.23   A) $C_4H_8N_2O_3(s) + 3\ O_2(g) \to (NH_2)_2CO(s) + 3\ CO_2(g) + 2\ H_2O(l)$

B)   80.3 J/K   C)   90.8 J/K

3.24   A)   Reaction I: 175 J/K   Reaction II: 538.4 J/K

B)   Reaction I: 176.4 J/KReaction II: 541.5 J/K

C)   The reactions mainly involve solids and liquids, not gases, so ΔS_r does not change markedly with small changes in T.

3.25 56.6 J/K mol

3.26 Answer summary:

	A (Isothermal)	B (Adiabatic)
w	−5710 J	0
q	+5710 J	0
ΔU	0	0
ΔH	0	0
ΔS	0	19.1 J/K

3.27 Answer Summary:

	STEP 1	STEP 2	STEP 3	TOTAL
w	0	5.19 kJ	−6.48 kJ	−1.35 kJ
q	−5.19 kJ	0	6.48 kJ	1.35 kJ
ΔU	−5.19 kJ	5.19 kJ	0	0
ΔH	−8.65 kJ	8.65 kJ	0	0
ΔS	0	14.4 J/K	−14.4 J/K	0

3.28 (A): Analysis:

(1)   Heat will be transferred, until the same final temperature is produced so that each block undergoes a ΔS due to ΔT.

(2)   The final temperature has to be determined by calorimetry starting with $-q_{\text{lost}} = q_{\text{gain}}$ before the ΔS can be calculated.

(3)   Then, when the results or T_f is substituted into the ΔS equation(s), the proof should be clearer.

A)   Detailed proof in Full Solutions

B)   (a)

Set	I	II	III
Block A	373 K	453 K	473 K
Block B	293 K	373 K	873 K
ΔS (J/mol K)	0.374	0.232	2.73

(b)   Set III should have greatest ΔS because ΔT is the largest.

(c)   Even though difference in T are the same for I and II, the blocks in II start at higher T's and already have higher S values.

3.29 See Full Solutions for full problem analysis.

$$\Delta S_{\text{total}} = \Delta S_{\text{cool gas}} + \Delta S_{\text{condense}} + \Delta S_{\text{heat liquid}}$$

$$= (-7.08 + -18.8 + 28.3)\ \text{J/K} = \underline{+2.42\ \text{J/K}}$$

3.30 See Full Solutions for full problem analysis.

A) Ice completely melts   B) 41.5°C   C) $\Delta S_{\text{ice}} = +170$, $\Delta S_{\text{warm water}} = -146$, $\Delta S_{\text{total}} = +24$ J/K

3.31 See Full Solutions for full problem analysis.

A) $T_{\text{final}} = 57.1°C$ B) $\Delta S_{\text{total}} = +27.8$ J/K

3.32 See Full Solutions for full problem analysis.

A) $P_{\text{total}} = 3.36$ atm

B) $\Delta S_{\text{mix}} = 5.56$ J/mol K

C)   Initial volumes and number moles of each gas are most important factors for ΔS_{mix}

3.33 See Full Solutions for full problem analysis.
  A)   5.6°C since solid and liquid are in equilibrium
  B) $\Delta S_{\text{total},C_6H_6} = \Delta S_{C_6H_6 \text{ cooling}} + \Delta S_{C_6H_6 \text{ freezing}} = -31.3 - 10 = \underline{-41.3} \text{ J/K}$

  C)   Yes, since some ice must have melted, even though equilibrium T maintained.
    ■ To calculate the moles of ice melted unless we apply calorimetry.
    ■ Assume q_{lost} benzene $= q_{\text{gain}}$ for the ice and only the phase change takes place.
    $\Delta S_{\text{ice}} = 42.5 \text{ J/K}$
  D) $\Delta S_{\text{surr}} = 1.2 \text{ J/K}$
  E)   (a) The $\Delta S_{C_6H_6}$ to become more negative, since more heat would be lost.
    (b) The ΔS_{ice} to increase, since more ice would melt and become more positive.
    (c) The difference between (a) and (b) will still be positive, since this would still be a spontaneous process.
    (d) *New values*: For C_6H_6: The value calculated for the liquid cooling stays the same, but the value for the phase change would change, since all benzene freezes and a third term $\Delta S_{(3)}$ for cooling the solid must be added.

$$\Delta S_{\text{total},C_6H_6} = \Delta S_{C_6H_6 \text{ (l)cooling}} + \Delta S_{C_6H_6 \text{ freezing}} + \Delta S_{C_6H_6 \text{ (s)cooling}}$$
$$= -42.7 - 10.0 - 2.6 = \underline{-55.3} \text{ J/K}$$

For ice in water: $\Delta S_{H_2O(s)} = \dfrac{q_{\text{gain}}}{T_{\text{mp}}} = \dfrac{1.55 \times 10^4 \text{ J}}{273 \text{ K}} = \underline{56.9} \text{ J/K}$

Process is still spontaneous because:

$$\Delta S_{\text{surroundings}} = \Delta S_{C_6H_6 \text{ total}} + \Delta S_{H_2O} = \left(-55.3 + 56.9\right)\dfrac{\text{J}}{\text{K}} = \underline{1.6} \text{ J/K}$$

3.34 A)   Derivation in Full Solutions
  B)   (a) 24.9 J/K mol   (b) For $CO_2(g)$: 25.6 J/K mol   For $CCl_4(g)$: 27.3 J/K mol

3.35 A) Derivation in Full Solutions
  B)   (a) 24.9 J/K mol   (b) For $CO_2(g)$: 23.2 J/K mol   For $CCl_4(g)$: 3.65 J/K mol
  C)   (a) Treating as a virial gas caused ΔS to decrease versus an ideal gas, while treating as a van der Waals gas produced increased values.
    (b) The second virial coefficient B takes into account both the "a" and "b" terms from the van der Waals equation.
    (c) The second virial coefficient B varies with temperature whereas both the "a" and "b" terms from the van der Waals equation do not.

PART 4: FREE ENERGY (ΔG), HELMHOLTZ ENERGY (ΔA), AND PHASE EQUILIBRIUM

4.1 A)   (a) T_1, T_2, and T_4 are single phases, only T_3 represents a phase transition.
    (b) At T_1 and T_2 the solid is most stable and at T_4 the gas state.
    (c) Gas + solid
  B)   At no T does the liquid represent the lowest μ, so that it is no transition to the liquid appearing for substance X. It will sublime before it melts.

4.2 A)   Circle: entropy "is positive" then enthalpy "is negative"
  B)   Circle: "increases"   C) $\Delta P = 0$ and $\Delta G = 0$

4.3 A)   26.8 J/K   B) −8.56 kJ   C) −8.56 kJ

4.4 A) ΔU, ΔH B) ΔS C) ΔU D) ΔG, ΔU E) none   F) none

4.5 A)   Need balanced reaction: $C_6H_6(l) + 7.5 O_2(g) \rightarrow 6 CO_2(g) + 3 H_2O(l)$
    ■ Maximum non-PV work means we need to calculate ΔG: (a) $\Delta G = -41.1$ kJ/g
  (b) Since ΔH and ΔS are the same signs, there will be a T at which ΔG can become positive, but $T_{\text{changeover}} = \dfrac{\Delta H}{\Delta S} = \dfrac{-3.203 \times 10^6 \text{ J}}{-218 \text{ J/K}} = 14{,}716 \text{ K}$ so it is not practical to

stop the reaction by raising T. The reaction won't stop (once started) until either all the fuel or the $O_2(g)$ is used up.

B) $H_2(g) + 1/2 O_2(g) \rightarrow H_2O(l)$ On a per gram basis: $\Delta G = -118.5$ kJ/g $H_2(g)$ has a much higher fuel value, that is nearly 3 times that of C_6H_6.

4.6 A) $\Delta S° = -48.4$ J/K $\Delta G° = 66.3$ kJ   B) $\Delta U° = 53.1$ kJ $\Delta A° = 67.5$ kJ

C) Both $\Delta G°$ and $\Delta A°$ indicate the reaction is non-spontaneous at all temperatures, since they will be positive values at all temperatures.

4.7 A) $\Delta H° = 128.3$ kJ $\Delta S° = 279$ J/K $\Delta G° = 45.2$ kJ

B) $\Delta U° = 133.2$ kJ $\Delta A° = 50.1$ kJ

C) Changeover T for $\Delta G° = 460$ K ($187°C$); for $\Delta A° = 477$ K ($204°C$)
– Not the same since $\Delta H \neq \Delta U$ for this reaction.

4.8 A) $\Delta H_r° = -1601$ kJ $\Delta S_r° = 4.67$ kJ/K $\Delta G_r° = -2993$ kJ

B) Since the sign of ΔH is negative, and ΔS positive, reaction will be spontaneously at all T's.

C) $\Delta U_r° = -1688$ kJ $\Delta A_r° = -3080$ kJ   D) About 2.9% more non-PV work as ΔA

4.9 A) $\Delta S = 7.81$ J/mol K

B) Since the signs of ΔH and ΔS are the same, T will definitely affect the direction of any transition.

C) $\Delta G = \Delta\mu = 0.105$ kJ at $0°C$ which indicates that the alpha phase has the LOWER μ value and it is the most stable, rather than the metallic beta phase.

D) Applying the equation for pressure dependence of a solid gives $\Delta G = \Delta\mu = -105$ J which indicates that now beta has the lower μ value and will be the stable phase with the increase in pressure at $13.2°C$.

4.10 A) (a) 22.4 kJ   (b) 22.4 kJ
(c) 1.78 kJ ■ Will need to calculate molar volume from the density of the liquid.

B) ΔG for CH_4 and C_2H_6 the same since both are gases (so the same equation applies) and we have the same number of moles of each. As long as gases are ideal, the ΔG with P is independent of chemical identity and only depends on the number of moles of gas.

C) C_6H_6 is a liquid and cannot substitute V_{gas} for ΔV_m. We will need actual density of liquid so chemical identity will affect the ΔG with P.

4.11 A) -195 atm

B) (a) -18.6 kJ
(b) The T_{mp} decreased by $3.5°C$ when P decreased. The T_{tr} from alpha to beta forms, although also decreased, would not have the same ΔT since the ΔV and ΔH_{tr} would be different values.

C) $\Delta S(\alpha \rightarrow \beta) = 443$ J/K mol, $\Delta S_{mp} = 587$ J/K mol
(a) The $\Delta S°$ for the $\alpha \rightarrow \beta$ transition in the fat involves about two-thirds of $\Delta S°$ for the melting, so the $\alpha \rightarrow \beta$ transition must involve a significant increase in disorder or freedom of movement for the chains.
(b) The $\alpha \rightarrow \beta$ transition in Sn results in a $\Delta S°$ 4.67 kJ/K mol, about 10 times the size of the $\Delta S°$ in the fat. The transition in the fat involves a shift of the chains and a small compression, whereas many more atoms have to change position in Sn, so it is consistent

4.12 A) $\Delta H = 544$ kJ/mol $\Delta S = 1.58$ J/K mol $\Delta G = 0$

B) At $37.0°C$: $\Delta H = 328$ kJ/mol $\Delta S = 917$ J/K mol $\Delta G = 43.1$ kJ/mol

C) Error, overestimate by 24% (calculate 50.kJ instead of 43.1 kJ/mol)
As illustrated in Part 2, since values for C_p's for protein folding are in kilojoules per mole instead of joules, they cannot be neglected.

D) $\Delta T = 0.003$ K. so no effect of changing pressure by 10 atm on T_m value.

4.13 A) $\Delta H°_{298} = 330$ J/mol $\Delta S°_{298} = 0.80$ J/K mol

B) $\Delta H°_{tr} = 398$ J/mol $\Delta S°_{tr} = 1.08$ J/K mol   C) 2.71 atm

4.14   Given that unfolded (α) $\rightarrow$ folded (β) is the direction, then the folded state (β) will be favored up until 377 K, above which the unfolded state (α) would be favored. [See diagram in Full Solutions]

4.15   A) $\Delta T = 0.53°C$ T_{fp} @ 100 atm $= -38.4°C$ B)   a 1.4% increase
    C)   The ΔT would have been calculated as negative ($-0.088°C$), indicating that the melting point was lowered. That can't be true since, when P increases, μ increases, and the intersection of μ_{solid} with μ_{liquid} should have been shifted up, not down. The ΔT calculated is only 16.7% of what it should be.

4.16   A) $P_{needed} = P_2 = 727$ atm (Using approximation equation, $P_2 = 719$ atm)
    B)   Skate produces $P = 132$ atm, so not enough ΔP to lower T_{mp} to $-5.0°C$.

4.17   A)   Modify Clapeyron Equation, $dT/dP = 28.0$ K/atm
    B) T_{bp} increases with P increase
    C)   Whenever T is in an equation, must use degrees Kelvin. Only allowed to use °C for ΔT values since amount of heat is the same whether expressed as ΔT(K) or ΔT(°C).
    D)   Change in boiling point is huge versus change in melting point with same ΔP. Two factors different: (1) $\Delta H_{vap} \gg \Delta H_{fus}$ but this would mean ΔT smaller since dividing by larger number if all else were the same and (2) difference in molar volume. The second factor is the more important in producing the increase in dT/dP.

4.18   A)   Because vapor pressure equals atmospheric pressure when a liquid boils, the water's boiling point will be below 100°C where $P = 1.0$ atm
    B)   Clausius Clapeyron equation since $\Delta V_m \approx V_{m,gas}$ approximation would apply.
    C)   354.4 K (81.3°C)

4.19   ■ Can use modified Clapeyron equation or Clausius Clapeyron equation because of the liquid-to-gas transition.
    T @ 10 psi $= 388$ K (115°C)

4.20   A) $\Delta H_{vap} = 33.1$ kJ/mol   B) $\Delta S_{vap} = 93.8$ J/K mol

4.21   A) $w = -31.0$ kJ $q = 40.66$ kJ $\Delta U = 9.6$ kJ $\Delta G = 0$ $\Delta S = 109$ J/K
    B) $\Delta G°(P_2) = 7.14$ kJ   Because $\Delta G°$ is positive, the liquid will be the stable state and conversion will not occur at 100°C.
    C)   If P is decreased, $\ln(P_2/P_1)$ becomes negative, so the conversion to the gas state becomes spontaneous at the lower P.

4.22   A) $\Delta G°_{298} = -724$ kJ/mol
    B)   Le Chatelier's principle would say that an increase in P would cause system to try to contract and favor the side with the fewer number of moles of gas. That means the ratio of reactant (pyruvate) to product (acetaldehyde) should increase and less CO_2(g) is produced.
    C) $\Delta G°_{298}$ (100 atm) $= -713$ kJ/mol which indicates that, since ΔG is less negative, less spontaneous, fewer products will be made at the higher P, as Le Chatelier's principle predicted.

4.23   A)   Vapor pressure $= 2.57$ atm   B) ΔG_{vap} @ 2.57 atm $= 2.35$ kJ
    Since ΔG for the vaporization is positive, the liquid is the more stable state, and will not all spontaneously convert to gas so there is no liquid in the lighter.

4.24 P needed $= 2.01$ torr

4.25   Vapor pressures at 25°C: ethanol 67.6 torr, acetone 236 torr and n-hexane 7.90 torr means the liquids could be identified by measuring the vapor pressures only.

4.26   A)   63.5 torr   B)   57.9 torr
    C)   The ΔC_p term does make a significant difference (due to the low ΔH_{vap}) so it should be included.

4.27 A) $y = \ln P$ $x = 1/T$

B) Plot $\ln P$ versus $1/T$, P must be in atm (so that $\ln P_1 = 0$) and T in Kelvin

C) $\Delta H_{vap} = -R$(slope) so $T_{bp}^\circ = -$(slope)$/b$

D) $\Delta H_{vap} = 45.1$ kJ (underestimate by 4.1%) $T_{bp}^\circ = 370$ K (which matches literature value)

E) Calculated $T_c = 509$ K (underestimate by 5.2%)

The assumption that ΔH_{vap} is constant from $T_1 \to T_2$ is probably no longer correct and results in the underestimation of T_c.

4.28 A) $\Delta H_{vap} = 29.4$ kJ/mol $T_{bp}^\circ = 316.7$ K (43.7°C)

B) (Including ΔC_p) $\Delta H_{vap} = 28.85$ kJ/mol $T_{bp}^\circ = 317.9$ K (44.8°C)

   ■ The inclusion of ΔC_p produces a result closest to the literature value, at 46.3°C

4.29 A) Derivation in Full Solutions

B) (a) $\Delta H_{vap} = 48.6$ kJ/mol $\Delta H_{sub} = 68.8$ kJ/mol   (b) $\Delta H_{fus} = 20.2$ kJ/mol
   (c) $T_{fp}^\circ = 489.5$ K (216.4°   C)   (d) $T_{triple} = 354$ K (81.0°C)

4.30 A) (1) (a)   A   (b) F, H   (c) C   (2) (a)   D   (b) E   (c)   G
   (3) B   (4) I   (5) C

B) Region (1): D   Region (2):   C   Region (3):   E   Region (4): H   Region (5): K

Point G	Point H	Point I	Point J
X exists as gas only	Two phases exist: X(l) ⇔ X(g)	Two phases exist: X(l) ⇔ X(g) But $d_{liquid} = d_{gas}$ so the meniscus is not visible	Single phase X(g) in supercritical region

4.31 A) P range $5.12 \to 72$ atm, T range: $-100 \to 50$°C

B) $31.2 \to 50$°C   C) $P = 5.11$ atm, $T = -56.4$°C

4.32 A) Phase rule: $F = C - P + 2$

Points	P	F	Phases in equilibrium	Single phase present
A	1	2		S (rhombic)
B	2	1	S (rhombic) + gas	
C	3	0	S (rhombic), S (monoclinic) + Liquid	
D	2	1	S (rhombic) + S (monoclinic)	
E	1	2		S (monoclinic)
F	2	1	S (monoclinic) + gas	
G	2	1	S (rhombic) + S (monoclinic)	
H	2	1	S (rhombic) + Liquid	
I	1	2		Liquid
J	2	1	S (monoclinic) + Liquid	
K	3	0	S (monoclinic) + Liquid + gas	
L	2	1	Liquid + gas	

B) Starting with an equilibrium between the two solid phases (rhombic, monoclinic), must change both P and T to maintain equilibrium (saying there is only one degree of freedom).

C) Increasing T but keeping P constant, $D \to J \to L$.

Point D	Observed as all solid, but would be equilibrium of the two crystalline forms, monoclinic and rhombic.
Point J	Increased T, some of the solid melted will have produced liquid, but there will be an equilibrium between monoclinic solid form and liquid.
Point L	All the solid would be melted, so would have liquid in equilibrium with gas.

D)   Decreasing P holding T constant, A → D → E → F.

At Point A	At Point D
μ(monoclinic) > μ(rhombic) So only rhombic form stable since it has lower μ value.	μ(monoclinic) = μ(rhombic) μ(monoclinic) decreased with P decrease so both crystalline forms are stable and coexist.
At Point E	**At Point F**
μ(monoclinic) < μ(rhombic) Decreased P has decreased μ monoclinic further so that it has become the most stable form.	μ(monoclinic) = μ(gas) μ(monoclinic) decreased with T increase and conversion from rhombic form has occurred.

E)   Point C at 95.4°C is a triple point in that there are three phases in equilibrium, with the two crystalline forms in equilibrium with the gas phase, so that μ(monoclinic) = μ(rhombic) = μ(gas). It is also the lowest temperature at which the rhombic form spontaneously converts to the monoclinic form, so it is considered to be the transition temperature. But 119°C would be considered the traditional triple point because μ(solid, monoclinic) = μ(liquid) = μ(gas).

F)   Point L at 445°C is when μ(liquid) = μ(gas) at 1.0 atm so it would be considered the normal boiling point for sulfur.

G)   The normal melting point would be the T at which μ(liquid) = μ(solid) when P = 1.0 atm and that is Point J on the diagram.

4.33 Answers:

Shifts	(1)	(2)	(3)	(4)	(5)
A)	isobaric	isothermal	Neither	isobaric	isobaric
B)	No change	No change	Liquid appears in the tube	No change	Both solid and liquid in tube
	(6)	**(7)**	**(8)**	**(9)**	**(10)**
A)	Neither	Neither	isothermal	Neither	isobaric
B)	Liquid disappears, only solid in tube	Solid + Liquid + gas in tube	Solid + liquid only in tube	Liquid + gas, but no solid in tube	No change

C)   (3), (5), (7), (8) and (9)

4.34 See phase diagram sketch in Full Solutions.

4.35 A)   **False** Don't need equal amounts of the two phases, just that the chemical potentials are the same. The key factors were ΔS_m and ΔV_m, as you can see from the map.

B)   **True**

C)   **False** The Clapeyron equation applies to the solid ⇔ liquid transition, not Clausius Clapeyron. The Clausius Clapeyron applies to a phase transition involving a gas as the final or initial state.

4.36 See Full Solutions

4.37 See Full Solutions.

4.38 A)   For 20 atm, 150°C: f = 19.3 atm (A 3.5% decrease)   B)   −119 J/mol
        For 200 atm, 150°C: f = 144 atm (A 28.2% decrease)   B)   −1188 J/mol

4.39 A)   See Full Solutions
        B)   Calculated answers:

van der Waals		P ideal = 10 atm			
T(°C)	ln ϕ	ϕ	Fugacity, ϕP	$\Delta\mu = RT$ ln ϕ	
25	−0.0522	0.9491	9.49	−129.5	J/mol
150	−0.0223	0.978	9.78	−78.4	J/mol
300	−0.0098	0.9903	9.9	−46.6	J/mol

Virial equation					
$T(°C)$	$\ln \phi$	ϕ	Fugacity, ϕP	$\Delta\mu = RT \ln \phi$	
25	−0.1083	0.8973	8.97	−268.5	J/mol
150	−0.0289	0.9716	9.72	−101.5	J/mol
300	−0.0097	0.9904	9.9	−46.2	J/mol

C) See Full Solutions

4.40 A) If B is negative, then the result of the integral will be negative, meaning ϕ will be less than 1.0 and the effective pressure will be less than the expected ideal gas pressure. When it becomes positive (350 K and above) the real pressure will exceed the ideal gas pressure.

B) Calculated results for B in table on the right:

	T (K)	B (L/mal)	P (atm)	$BP/RT = \ln \Phi$	$\Delta\mu = RT \ln \Phi$		Φ
	200.0	−0.0355	100	−0.2166	−360.1	J/mol	0.8053
	273.0	−0.0108	100	−0.0484	−109.8	J/mol	0.9528
	350.0	0.00270	100	0.0094	27.4	J/mol	1.0094
	450.0	0.0127	100	Q.03440	128.7	J/mol	1.0350
	550.0	0.0188	100	Q.Q4162	190.3	J/mol	1.0425
C)	T (K)	B (L/mol)	P (atm)	$BP/RT = \ln \Phi$	$\Delta\mu = RT \ln \Phi$		Φ
	200.0	−0.0355	20	−0.0433	−72	J/mol	0.9576
	273.0	−0.0108	20	−0.0097	−22	J/mol	0.9904
	350.0	0.00270	20	O.OOLB8	5.5	J/mol	1.0019
	450.0	0.0127	20	0.00638	25.7	J/mol	1.0069
	550.0	0.0188	20	Q.OOS32	1	J/mol	1.0084
	T (K)	B (L/mol)	P (atm)	$BP/RT = \ln \Phi$	$\Delta\mu = RT \ln \Phi$		Φ
	200.0	−0.0355	2	−0.0043	−7.2	J/mol	0.9957
	273.0	−0.0108	2	−0.001	−2.2	J/mol	0.9990
	350.0	0.00270	2	0.00019	0.5	J/mol	1.0002
	450.0	0.0127	2	0.00069	2.6	J/mol	1.0007
	550.0	0.0188	2	0.00083	3.8	J/mol	1.0008

C) (a) The $\Delta\mu$ values are getting smaller as the P is lowered.

(b)

P	$\mu_{real} > \mu_{ideal}$	$\mu_{real} < \mu_{ideal}$	$\mu_{real} = \mu_{ideal}$
100	350 → 550 K	200, 273 K	None listed
20	450 → 550 K	200, 273 K	≈350 K
2	None listed	200 K	≈273 → 550 K

(c) The T regions where the attractive and repulsive forces dominate definitely change with P.

T	$P = 100$ atm	$P = 20$ atm	$P = 2.0$ atm
200 K	Attractive forces	Attractive forces	Attractive forces
273 K	Attractive forces	Attractive forces	Neither dominating
350 K	Attractive forces	Neither dominating	Neither dominating
450 K	Repulsive forces	Repulsive forces	Neither dominating
550 K	Repulsive forces	Repulsive forces	Neither dominating

(d) Yes, the behavior observed with the P decrease fits the figure shown. However, depending on the temperature, when pressure is constant, the μ_{real} could also be greater than, equal to or less than μ_{ideal}, for a range of temperatures. So a plot of μ versus T(K) would look somewhat similar to the μ versus P plot, except the $\mu_{real} > \mu_{ideal}$ values will appear at low temperatures, whereas they should appear at higher pressures as in the figure shown. Increasing T increases kinetic energy and makes it harder for the molecules to cluster.

PART 5: FREE ENERGY (ΔG) OF MIXING, BINARY LIQUID MIXTURES, COLLIGATIVE PROPERTIES, AND ACTIVITY

5.1 A)   (a) $\Delta G_{mix} = -1.70$ kJ/mol   (b) $\Delta S_{mix} = 5.76$ J/mol K

    B)   If P decreases but the composition stays the same, ΔG_{mix} and ΔS_{mix} stay the same.

5.2 A) $\chi_{CO_2} = 0.0909$, $\chi_{O_2} = \chi_{He} = 0.4545$

    B) $n_{total} = 1.80$ mol $\Delta G_{mix} = -4.16$ kJ

5.3 A)   16.18 L of $O_2(g)$

    B)   (a)   The ΔG_{mix} should become more negative since mixture getting closer to equal molar mixture.

          (b) ΔG_{mix} air $= -1273$ J/mol, ΔG_{mix} Nitrox I $= -1553$ J/mol, so more negative by -230 J/mol

5.4 A)   1048.0 kPa (10.35 atm)

    B) $P_{He} = 628.6$ kPa (6.21 atm) $P_{N2} = 300.6$ kPa (2.97 atm)

    C) -2.26 kJ/mol

5.5 A) $\Delta G_{mix} = -1.59$ kJ/mol $\Delta S_{mix} = 5.35$ J/K mol

    B)   (a)   Because the number of moles of each gas changes, the values for ΔG_{mix} and ΔS_{mix} will have to change since the mole fractions change.

       ■   Must calculate number of moles of each gas from the initial conditions. The new values are $\Delta G_{mix} = -1.39$ kJ/mol and $\Delta S_{mix} = 4.67$ J/K mol

          (b)   The ΔG_{mix} will only be part of ΔG_{total} since the pressure changes for each gas will contribute to ΔG_{total}.

       ■   Must determine the final pressure of the gas mixture to calculate ΔG for pressure changes.

       ■   Because ΔG is a state function, you can add the separate changes together to get ΔG_{total}.

$$\Delta G_{total} = \Delta G_{mix} + \Delta G_{N_2, \Delta P} + \Delta G_{Xe, \Delta P}$$

    C) $= \left[0.6543 \text{ mol} \left(-1393 \dfrac{J}{mol} \right) \right] + (-271 \text{ J}) + (190) = \underline{\underline{-992 \text{ J}}}$

       ■   Because ΔG_{total} is negative, the mixing and pressure changes will occur spontaneously.

5.6 A)   Need 1 mol C_6H_{14} for every 1 mol C_7H_{16} in the mixture.

    B)   46.2% by mass C_6H_{14}, 53.7% by mass C_7H_{16}

    C)   118.3 mL C_6H_{14} and 131.7 mL C_7H_{16}

5.7 A)   Water is the solvent and at $\chi_B = \approx 0.30$ there are 7 water molecules for every 3 CH_3OH molecules. The contraction is because the methyl group of CH_3OH cannot H-bond so it is disrupting water's organization and causing water's structure to partially collapse on itself.

    B)   (a)   Pure molar volumes: $\bar{V}_A^* = 18.04 \dfrac{cm^3}{mol}$; $\bar{V}_B^* = 40.703 \dfrac{cm^3}{mol}$

          (b)   From graph: $\Delta V_A = -0.35 \dfrac{cm^3}{mol}$ and $\Delta V_B = -2.0 \dfrac{cm^3}{mol}$ so that

$$\bar{V}_A = 18.04 + (-0.35) = 17.7 \dfrac{cm^3}{mol} \text{ and } \bar{V}_B = 40.703 + (-2.0) = 38.7 \dfrac{cm^3}{mol}$$

       ■   Both agree with open circle values on second graph.

    C)   (a)   Approximately 0.90–0.95 cm³/mol

          (b)   At χ $CH_3OH = 0.70$, $\Delta V_A = -2.0 \dfrac{cm^3}{mol}$ and $\Delta V_B = -0.40 \dfrac{cm^3}{mol}$ so that

          (c) $\bar{V}_A = 18.04 + (-2.0) = 16.0 \dfrac{cm^3}{mol}$ and $\bar{V}_B = 40.703 + (-0.40) = 40.3 \dfrac{cm^3}{mol}$

       ■   Both agree well with red circle values on second graph.

5.8 A) (a) The solution is contracting.

(b) Expect that, since both water (H_2O) and isopropanol ($CH_3CHOHCH_3$) have H-bonding as intermolecular forces, the A⋯B interactions would be similar in strength to A⋯A and B⋯B interactions and the two substances should form a mixture. The ΔG_{mix} overall should then be negative so that mixing is spontaneous.

(c) Since isopropanol has only one OH group centered between bulky methyl groups, the H-bonding network in water is being disrupted by the alcohol molecules (as in the CH_3OH and C_2H_5OH mixtures shown previously). Then as the mole fraction of the alcohol increases, the water molecules between the alcohol molecules allow them to be closer together than they would be in pure isopropanol.

B) (a) Need to calculate pure molar volume for isopropanol since:
$$\overline{V}_{ideal} = \chi_{H_2O}\overline{V}^*_{H_2O} + \chi_{isop}\overline{V}^*_{isop}$$

$$\overline{V}^*_{isop} = \frac{60.096\ \dfrac{g}{mol}}{0.7804\ \dfrac{g}{cm^3}} = 77.01\ \frac{cm^3}{mol} \quad \text{then: } \overline{V}_{ideal} = \underline{\underline{35.71}}\ \frac{cm^3}{mol}$$

(b) $\overline{V}_{soln} = \overline{V}_{ideal} + \Delta\overline{V}_{mix} = \underline{\underline{34.93}}\ \dfrac{cm^3}{mol}$ (c) $\overline{V}_{isop} = \dfrac{V_{soln} - \chi_{H_2O}\overline{V}_{H_2O}}{\chi_{isop}} = \underline{\underline{75.83}}\ \dfrac{cm^3}{mol}$

5.9 A) $\overline{V}^*_{H_2O} = 18.05\ \dfrac{cm^3}{mol} \quad \overline{V}^*_{C_2H_5O} = 58.37\ \dfrac{cm^3}{mol}$

B) (a) H_2O: $\chi_{C_2H_5O} = 0.0 \rightarrow 0.10$ (b) At NO mol fraction

C) (a) H-bonding (b) No, should stay the same

D) The ethanol molecules are able to insert themselves into water's H-bonding network initially, causing expansion, but because there are only 2 H-bonding sites on ethanol, instead of the 4 on water, increasing the number of ethanol molecules increasingly causes a collapse of water's 3-dimensional network of H-bonding.

5.10 A) $\Delta V_{ethanol} = 107.8$ mL (cm^3)

B) $\Delta V_{mix} = -5.14$ mL (cm^3) $\rightarrow -1.11\ cm^3/mol$

C) Yes. Graph shows value close to $-1.1\ cm^3/mol$

D) Yes, because the contraction is -5.1 mL below expected total (157.7 mL) should be able to detect in this way.

5.11 A) ▪ Use density and total weight of solution to get actual volume: $V_{soln} = 137.22\ cm^3$

B) (a) $\overline{V}^*_{CH_3OH} = 40.70\ \dfrac{cm^3}{mol} \quad \overline{V}^*_{(CH_3)_2CHOH} = 77.01\ \dfrac{cm^3}{mol}$ (b) $137.33\ cm^3$

C) (a) Ideal density $= 0.7845\ g/cm^3$, actual density is slightly greater at $0.7851\ g/cm^3$

(b) Solution has contracted, since there is more mass per unit volume than expected.

(c) Would have LDF interactions between the methyl groups on both molecules and an OH group for H-bonding so forces are very similar. The solution is very close to ideal since the contraction is very small. The different shapes of the molecules probably have the largest effect and cause the "normal" H-bonding and LDF networks in the pure liquids to be disrupted.

5.12 A) (a) H-bonding (b) Dipolar forces

(c) Dipolar forces, unless the water molecules H-bond to the O in the C=O bond.

B) (a) It is contracting, so $V_{soln} < V_{ideal}$ at all mole fractions of acetone.

(b) Six molecules of H_2O for every 4 molecules of $(CH_3)_2CO$ (or 3:2)

C) (a) water (b) At about 0.20

(c) That the acetone⋯water forces are stronger than the forces in pure water, so energy is released. It is consistent with some form of a H-bonded complex cluster (such as that in figure on the right) occurring between water and acetone, and the mole fraction of acetone is low.

D)   (a)   acetone

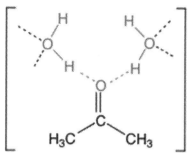

(b)   That the acetone⋯water forces are weaker than the forces in pure acetone so acetone largely interacts with itself in the solution arrangements. This is consistent since if you need more water molecules than acetone to form the cluster and at high mole fraction acetone, there are too few water molecules to do that.

5.13 A)   (a) Dipolar forces in pure chloroform   (b) Dipolar forces in pure acetone

B)   (a)   It contracts up to about $\chi_{acetone} \approx 0.7$, then begins to expand, versus an ideal solution.

(b)   The maximum occurs near $\chi_{acetone} \approx 0.5$ (7 chloroform: 3 acetone molecules).

C)   (a)   When the ratio is about 1:1 or $\chi_{acetone} \approx 0.5$.

(b)   The chloroform-acetone interaction is stronger than either chloroform with itself or acetone with itself.

D)   (a)   Since H-bonding interactions are stronger than dipolar forces, energy would be released and produce the exothermic ΔH_{mix} values, if they are being formed in the solution. The maximum loss should appear at a molar ratio of 1:1 and the data shows that.

(b)   It would be unusual since H's bonded to C do not participate in H-bonding. Only H's bonded to O or N are generally involved, so this would be a unique situation.

5.14 A) $\Delta G_{mix,ideal} = -1861$ J/mol $T\Delta S_{mix,ideal} = +1861$ J/mol

B)   Mixture I: $G^E = 250$ J/mol, $\Delta G_{mix} = -1611$ J/mol
Mixture III: $G^E = -500$ J/mol, $\Delta G_{mix} = -2361$ J/mol
Mixture IV: $G^E = 1500$ J/mol, $\Delta G_{mix} = -361$ J/mol

C)   See Full Solutions.

(b) All three mixtures have negative values for ΔG_{mix} so the two liquids will mix, although mixture IV will be the least spontaneous with a $\Delta G_{mix} < \Delta G_{ideal}$.

5.15 A)   **For the acetone, $(CH_3)_2CO/H_2O$ mixture:** As the temperature is increased, the H^E becomes less endothermic at low mole fraction of acetone and more exothermic at higher mole fractions. The transition occurs at lower mole fractions at higher temperature.

For the acetonitrile, CH_3CN/H_2O mixture: As the temperature is increased, the H^E becomes more endothermic. Both changes are consistent with a positive term being added to H^E as the T is increased.

B)   Kirchhoff's Law: $\Delta H(T_2) = \Delta H(T_1) + \int_{T_1}^{T_2} \overline{C_p}\, dT$

C)   The heat capacity of the mixture at each composition from: $H^E(T_2) = H^E(T_1) + \overline{C}_{mix}\Delta T$.

5.16 A)   The P and T dependence of G (or ΔG): $\left[\dfrac{d\overline{G}}{dP}\right]_T = -\overline{V}$ and $\left[\dfrac{d(G/T)}{dT}\right] = -\dfrac{H}{T^2}$

B)   It changes the units on both the right and left sides to K^{-1} that can be easier to apply or plot.

C)   The conversion factor 101.3 J/L atm is needed for V^E.

D) $\dfrac{V^E}{RT} = \underline{6.13 \times 10^{-6}}\, \text{atm}^{-1}$ $\left[\dfrac{d(G^E/RT)}{dT}\right] = \underline{-0.0214}\, \text{K}^{-1}$

Since the volumes of liquids are generally not influenced by pressure changes, it is not surprising that the T dependence of G^E is much greater than the P dependence.

5.17   A)   Both molecules are weakly polar and would have dipolar forces so that $A{\cdots}B \approx A{\cdots}A \approx B{\cdots}B$.

B) $P_{total} = 96.5$ torr   C) $y\ C_4H_9Br = 0.219$ D) -1.70 kJ/mol

5.18   A)   Both molecules have the same basic shape (tetrahedral) and similar size so that the contact area (which affects LDF) should be about the same. $CHCl_3$ is weakly polar, whereas CCl_4 is non-polar, but expect $A{\cdots}B$ as LDF would be very close to the forces with the pure components.

B)   ▪ Use the liquid composition and Raoult's law to get partial pressure for each component, P_{total} and then calculate the mole fractions in vapor. Given: $P^*_{CHCl_3} = 26.54$ kPa and $P^*_{CCl_4} = 15.27$ kPa

$$P_{total} = \chi_{CHCl_3} P^*_{CHCl_3} + \chi_{CCl_4} P^*_{CCl_4} = 0.667(26.54\,kPa) + 0.403(15.27\,kPa)$$

$$= 17.70 + 5.09 = 22.79\,kPa$$

Then: $y_{CHCl_3} = \underline{0.777}$ and $y_{CCl_4} = 1.0 - 0.777 = \underline{0.223}$

C)   Molar ratio in vapor is 0.777 $CHCl_3$:0.223 mol $\overline{CCl_4}$ = 3.5:1.0 or **7 mol $CHCl_3$ for every 2 mol CCl_4** which is much higher than the 2:1 ratio in the liquid.

D) $P_{atmosphere}$ lowered to 22.79 kPa or 171 torr.

5.19   A) $\chi_{hexane} = 0.210$ B) $y_{hexane} = 0.580\ y_{octane} = 0.420$

C)   (a) See Full Solutions   (b) ΔG_{mix} liquid $= -1.59$ kJ/mol and ΔG_{mix} vapor $= -2.11$ kJ/mol

▪ Should not be the same ΔG_{mix} since the mole fractions of components change.

5.20   A)   See the Full Solutions for proof.

B)   (a) $P^*_{CCl_4} = \underline{6.08\ atm}$ (b) $y_A = \dfrac{P^*_A P_{total} - P^*_A P^*_B}{P_{total}\left(P^*_A - P^*_B\right)} = \underline{0.667}$

(c) %difference $= \underline{5.7\%}$ (d) 0.9460

5.21   A)   (a) Raoult's law   (b) P_B (c) 7 (and possibly 10)   (d) (1), (4)

(e) (1) (and possibly 2)   (f) P_{Total}, ideal   (g) Actual P_{Total} (h) P^*_B

B)   (a) and (d)

5.22   A)   Since negative deviations seen, G^E and H^E must be exothermic values.

B)   It is likely there would be a maximum azeotrope in the T–xy diagram for this mixture since the deviations are negative.

C)   Component A, because the mole fraction in the vapor at this point is 0.717 from estimated values of P_A and P_{total} from the graph (or because $P^*_A > P^*_B$).

D) $K_{H,B} < P^*_B$ since negative deviation produce an intersection below the pure vapor pressure.

5.23   A)   Diethyl ether ($CH_3CH_2OCH_2CH_3$) has dipolar forces, acetone (CH_3OCH_3) also has dipolar forces.

B)   (a) See calculated data and plot in Full Solutions.

(b) Positive deviations occur, so not acting ideally.

(c) $\gamma_{acetone} = 1.19\ \gamma_{diethyl\ ether} = 1.21$

C)   At nearly pure diethyl ether, $\chi \approx 0.90$ the mole fraction of diethyl ether in the vapor equals that in the vapor. See calculated data and plot in Full Solutions.

5.24 A) The pure vapor pressure of both liquids increases as T increases. This behavior should then be true of all liquid mixtures.

B) Yes the basic shape is retained and the curves come together at the same mole fraction (≈ 0.90) at each temperature.

C) (a) $P = 64$ kPa, 303.15 K: $\chi_{diethylether} = 0.30$ $y_{diethylether} = 0.58$

(b) $P = 36$ kPa, 293.15 K: $\chi_{diethylether} = 0.15$ $y_{diethylether} = 0.40$

(c) $P = 20$ kPa, 273.15 K: $\chi_{diethylether} = 0.50$ $y_{diethylether} = 0.65$

D) For two reasons:

(1) The difference between the pure vapor pressures is decreasing. [*As can be seen on the graph at mol fractions 0 and 1.0*].

(2) As T increases the vapor phase becomes "richer" in the more volatile component and the curves become more separated. [*As can be seen by the difference between the values in the figure shown in the problem*].

5.25 A) See Full Solutions for graph

B) Answers: (a) Positive deviations

(b) $CHCl_3$ shows much greater positive deviations than ethanol. That shows that the ethanol–$CHCl_3$ forces are much weaker than the pure ethanol forces, or $CHCl_3$ forces, since $CHCl_3$ escapes much more easily from the mixture than what we would expect (ideal value) if it experienced the same forces.

(c) The activity coefficient (γ) is always greater than 1.0 since P_{obs}, $CHCl_3$ is greater than P_{ideal}, $CHCl_3$ at all mole fractions.

(d) The value for γ C_2H_5OH would be close to 1.0 from χ C_2H_5OH $0.60 \rightarrow 1.0$, but is greater than 1.0 at mol fractions C_2H_5OH below 0.40.

5.26 A) Yes, positive deviations indicate the A$\cdots$B interactions are weaker than what the pure liquids have and that should indicate a positive value for G^E, so it is consistent.

B) The log of the activity coefficient for ethanol starts out above 1.0, but then drops to lower values, indicating the interactions of ethanol and $CHCl_3$ molecules start out as very weak. It takes added energy to replace the ethanol H-bonds with the dipolar ethanol–$CHCl_3$ interactions and H^E is endothermic. However, as ethanol's mole fraction increases, ethanol can re-establish H-bonding and the H^E becomes exothermic.

C) The activity coefficient for $CHCl_3$ quickly increases and doesn't show a region where it is close to 1.0, where the log γ would be zero. That correlates to the deviations seen on the Raoult's law diagram at nearly all mole fractions.

5.27 A) Yes, it has an azeotrope. The plot shows a minimum T_{bp} for the mixture, below either of the boiling points of the pure components. If the deviations from Raoult's law are positive, then a minimum boiling temperature will be observed so the T–xy plot shows the expected behavior.

B) Because the highest vapor pressure will produce the lowest boiling point for the mixture.

C) Yes, it does. The equality occurs at ≈ 0.85 mol fraction $CHCl_3$.

D) χ $CHCl_3 = 0.40$, y $CHCl_3 = 0.60$.

E) ■ Apply the lever rule.
The ratio is about 4:3 or 1.33 more moles $CHCl_3$ in liquid phase than in vapor.

5.28 A) Positive deviations and would expect only LDF forces between acetone and cyclohexane, since acetone has dipolar forces but cyclohexane, being non-polar, has only LDF forces.

B) Yes, there will an azeotrope and it will be a minimum T_{bp}, since positive deviations from ideal behavior occur in the vapor pressure.

C)

Point	A	B	C	D
	Only liquid mixture present (single phase) χ acetone = 0.30	Liquid + vapor phases in equilibrium χ acetone = 0.12 y acetone = 0.52	Vapor in equilibrium with a drop (small amount) of liquid χ acetone = 0.30 y acetone = 0.30	Vapor only (single phase) y acetone = 0.30

D) ■ Apply the lever rule. The ratio is about 1.1

5.29 A) Raoult's law Diagram – see Full Solutions.

B) Since both molecules are polar, A···B should be dipolar forces. A···B must be a slightly weaker interaction since the positive deviations are small but do occur.

C) (a) ■ Plot the partial pressure of iodoethane versus mole fraction iodoethane for the Henry's law region (first 5 values). $K_H\, C_2H_5I = 449$ torr.

(b) ■ Calculate the mole fraction of $C_4H_8O_2$ in the Henry's law region for ethyl acetate (χ iodoethane = 0.65 → 0.0) and plot versus partial pressure of ethyl acetate.
$K_H\, C_4H_8O_2 = 337$ torr

D) ■ Activity coefficients obtained by comparing the observed partial pressure to the ideal value.

$\chi\, C_2H_5I = 0.2353$ $\gamma\, C_2H_5I = 1.27$ $\gamma\, C_4H_8O_2 = 1.04$
$\chi\, C_2H_5I = 0.5473$ $\gamma\, C_2H_5I = 1.10$ $\gamma\, C_4H_8O_2 = 1.14$
$\chi\, C_2H_5I = 0.8253$ $\gamma\, C_2H_5I = 1.02$ $\gamma\, C_4H_8O_2 = 1.36$

5.30 A) Positive deviations   B) $\chi_{CH_3OH} = 0.505$

5.31 A) (a) Yes, the deviations are consistent since the excess properties H^E and G^E had indicated the A···B forces were stronger than those in pure acetone or $CHCl_3$. There is nothing to indicate in the Raoult's law diagram that a complex is being formed. We need properties like the excess thermodynamic properties to substantiate that.

(b) The activity coefficient, measured by the separation of the ideal partial pressure and actual partial pressure, seems to occur at $\chi\, CHCl_3 = 0.60$, where it is approximately 0.78.

(c) It is approximately 21 kPa.

B) (a) Yes, there should be a maximum boiling point since the deviations are negative, from Raoult's law.

(b) $\chi\, CHCl_3 = 0.60$ (c) $\chi\, CHCl_3 = 0.25$, $y\, CHCl_3 = 0.20$

5.32 A) (a) $P_{total,actual} > P_{total,ideal}$.

(b) K_H for ethanol $> P_{C_2H_5OH}^*$.

(c) Pure C_2H_5OH would have H-bonding forces whereas DIPE, $(CH_3)_2COC(CH_3)_2$, will have dipolar forces, but not H-bonding. The A···B forces will then be largely dipolar, which is weaker than the H-bonding between C_2H_5OH molecules.

B) (a) Yes, it will. (The relationship is illustrated in Problem 5.27)

(b) It will be a minimum T_{bp} since the deviations are positive (opposite behavior than the P–xy graph).

(c) It will occur at $\chi_{C_2H_5OH} = 0.40$

C) (a) Final answers:

(b) Yes, the sign for G^E mixture is positive throughout all mole fractions. This is what we would expect, that overall energy must be added to sever the stronger forces in the pure components to

x1	G^E ethanol [J/mol]	G^E DIPE [J/mol]	G^E mixture [J/mol]
0.198	2566.3	146.07	625.3
0.414	1260.2	761.87	968.2
0.618	512.7	1520.88	897.8
0.813	104.3	2503.29	552.9

form the weaker interaction between ethanol and DIPE, based on the thermodynamics of the interactions, and since the observed deviations are positive.

5.33 A) ▪ Need to look up ΔH_{vap} and normal T_{bp} for Br_2 liquid. $P^{\bullet}_{298} = 187\,torr$

B) See Full Solutions for proof. Positive deviations occur, indicating $Br_2 \cdots CCl_4$ interactions weaker than $CCl_4 \cdots CCl_4$ intermolecular forces.

C) $K_H = 414\,torr$

5.34 A) See the Full Solutions for calculated values.

B) Positive deviations occur in the mixture.

C) (a) See the Full Solutions for plot.

(b) The plot shown is very similar to that for di-isopropyl ether mixed with ethanol in Problem 5.32. Since both are ether/alcohol mixtures, we could expect that the disruptions caused by the formation of A···B would be very similar, since the H-bonding of the alcohol is most affected.

(c) Since the logs of the activity coefficients are nearly the same at $X = 0.50$, we would expect they both contribute about one-half of the G^E for the mixture.

D) (a) + (b) See the Full Solutions for calculated values.

(c) The two components will spontaneously mix at all mole fractions. The least spontaneous is a small amount of methanol is being added to the ether.

5.35 A) No azeotrope is visible so it is called a "zeotropic" diagram.

B)

Point	A	B	C	D
	Only vapor mixture present (single phase)	Liquid + vapor phases in equilibrium	Liquid + vapor phases in equilibrium	Liquid mixture only (single phase)
		χ hexane = 0.20	χ hexane = 0.30	χ hexane = 0.68
	y hexane ≈ 0.45	y hexane = 0.75	y hexane = 0.30	

C) The vapor pressure of hexane is much higher than that of m-xylene, indicating it has much weaker forces to overcome, so that, unless it has a very small mole fraction, it is always going to produce a higher proportion of the total pressure over the solution.

D) It would not be very easy to predict what the ideal boiling point should be for each mixture, to gauge the deviation, since both components are volatile and contribute to the total vapor pressure, so the P–xy graphs are generally used for that determination.

5.36 218 g/mol

5.37 ▪ Must calculate the mole fraction of $Ba(NO_3)_2$ in the solution.

▪ Apply $i_{NaCl} = \dfrac{\Delta P_{obs}}{\Delta P_{i=1.0}}$ A) 2.73   B) 91.0%

5.38 ▪ Must calculate the van't Hoff factor and apply to the mole fraction of NaCl in the solution.

Weight needed = 156 g NaCl

5.39 MW = 844 g/mol

5.40 A) $\chi_{C_3H_{18}O_6} = 0.187$

B) (a) $\Delta T_{fp} = -\dfrac{R(T^{\bullet}_{fp,H_2O})^2}{\Delta H_{fus,H_2O}}\chi_B = -19.3\,K = \underline{-19.3°C}$

(b) $\Delta T_{fp} = -K_{f,H_2O}(m_B) = \underline{-23.8°C}$

(c) The values don't agree, with the approximation producing the larger value.

It is likely that the approximation that $\chi_B \approx \dfrac{n_B}{n_A}$ doesn't apply, since that value equals 0.230, not 0.187.

5.41 A)   Molar mass = 139 g/mol   B) T_{bp} solution = 81.87°C

5.42 T_{bp} solution = 62.7°C

5.43 A)   Because the value for K_f for A depends on the normal freezing point of pure A, the ΔH_{fus} of A and its molecular weight, while K_b depends on the normal boiling point of pure A, the ΔH_{vap} of A and also the molecular weight of A.

B)   The K_f should be larger than K_b for the same solvent because, although the boiling point is a higher temperature than the freezing point, the division by the much larger ΔH_{vap} always makes the K_b smaller.

5.44 $K_{f,A} = \dfrac{R(T^{*}_{fp,A})^2}{\Delta H_{fus,A}}\left(\dfrac{1000}{MW_A}\right) = \underline{\underline{5.32}}\ \dfrac{K(°C)\text{-kg}}{mol}$

5.45 $\Delta H_{fus,A} = \dfrac{R(T^{*}_{fp,A})^2}{K_{f,A}}\left(\dfrac{1000}{MW_A}\right) = 8.67\ \dfrac{kJ}{mol}$

5.46 A) T_{bp} solution = 101.8°C   B) 73.4 atm

C)   If urea forms a dimer, then the effective number of particles decreases which would produce lower values for the boiling point elevation and in osmosis. The most noticeable change of a small decrease in moles will be the change in the osmotic pressure.

5.47 A)   Results:

Salt solution	MW (g/mol)	$\Delta T_{fp,obs}$	m (mol/kg)	ΔT_{fp} when i=1.0	Calc'd i	No. ions salt	% dev in i
14.0% NaCl	58.5	−9.94°C	2.785	−2.544	1.92	2	4.0%
14.0% MgSO₄	120.4	−2.86°C	1.352	−2.515	1.14	2	43%
14.0% BaCl₂	208.4	−3.92°C	0.7811	−1.453	2.70	3	10%
14.0% (NH₄)₂SO₄	132.1	−4.07°C	1.232	−2.291	1.78	3	40.7%
14.0% AgNO₃	169.9	−2.55°C	0.958	−1.782	1.43	2	28.5%

B)   Smallest % dev → largest % dev: $NaCl < BaCl_2 < AgNO_3 < (NH_4)_2SO_4 < MgSO_4$

5.48 A) 50.31 atm   B) If $I = 3.0$, overestimate osmotic pressure by 11.1%

5.49 1.522 M glucose

5.50 A)   0.300 M sucrose

B)   The χ_{H_2O} is lower inside the cell (where the solute concentration is higher) than in the D5W solution, so water will flow into the cell.

C) $i = 1.85$

D)   Yes, the NaCl data gives $i_{NaCl} = \dfrac{\Delta T_{obs}}{\Delta T_{i=1}} = 1.83$ so the van't Hoff factor can be below 2.0 even in low concentrations of solute and should be taken as 1.85 in (A), making $C_{sucrose} = 0.278$ M

5.51 A)   29.6 atm

B) $\Delta\Pi = \Pi_{mannitol} - \Pi_{D5W} = 29.6\ \text{atm} - 7.07\ \text{atm} = \underline{\underline{22.5\ \text{atm}}}$.

5.52 A)   See the Full Solutions for derivation: $\dfrac{\chi_{H_2O\ vitreous\ humor}}{\chi_{H_2O,mannitol}} = \exp\left[\dfrac{\Delta\Pi \bar{V}_{H_2O}}{RT}\right]$

B) $\dfrac{\chi_{H_2O\ vitreous\ humor}}{\chi_{H_2O,mannitol}} = 1.016 = 1.02$

C)   A mole fraction difference of 2% produces a 22 atm change in pressure, so even much smaller changes can be measured easily from osmotic pressure.

5.53 A)   MW = 74.1 g/mol

B)   Only the mole fraction of A is involved in the original derivation so that no properties specific to water are included in the final equations. Therefore the solvent can be any liquid as long as the membrane is permeable to only that liquid.

5.54 A) $MW_{creatine} = \underline{142.9} g/mol$

B) Given $P^*_{H_2O} @ 37°C = 47.067 \, torr$, and creatine a non-volatile solute, then:

$\Delta P = \chi_{creatine} P^*_{H_2O}$

Since $\chi_{creatine} = 1.26 \times 10^{-5}$, $\Delta P = 1.26 \times 10^{-5}(47.067 \, torr) = 5.93 \times 10^{-4} \, torr$

This pressure difference is too small to measure accurately, so that applying Raoult's law for a molecular weight determination won't be practical.

C) $20 \, mg/L \rightarrow \Pi = 3.423 \times 10^{-3} \, atm \left(\dfrac{760 \, torr}{1 \, atm} \right) = 2.60 \, torr$ and $750 \, mg/L \rightarrow \Pi = 97.5 \, torr$

The osmotic pressures could easily be measured accurately, in either torr or kPa, for these concentrations.

5.55 A) $5.17 \times 10^{-4} \, M$ B) 108.6 g polymer per kg water

5.56 A) *See the Full Solution for derivation*

Plot $y = \Pi, x = \left(\dfrac{mass \, B, g}{V_{soln}(L)} \right)$ then slope $= \dfrac{RT}{MW_B}; b \approx 0$

B) $3.82 \times 10^5 \, g/mol$

5.57 ▪ Change pressures to atm from torr before plotting data. MW = 53,670 g/mol

5.58 A) (a) 7.29 atm   (b) 7.31 atm   (c) 7.22 atm

B) The approximation that $\chi_B = n_B/n_A$

5.59 ▪ Get the concentration of solute from first property, then convert to needed units for second colligative property. T_{fp} solution $= -7.54°C$

5.60 A) ▪ Need to determine *i* factor for NaCl by comparing osmosis data.

▪ Must have concentration units in molarity before comparing osmotic pressures

For the NaCl solution $T_{fp} = -0.39°C$ B) Yes, for the mannitol solution $T_{fp} = -0.40°C$

5.61 A) $\Pi_{albumin} = 11.2 \, torr$, $\Pi_{globulin} = 2.42 \, torr$ B) 395 times greater

C) We can assume each Π value is independent of any other, since, as a colligative property, it doesn't matter what the solute particle is, but how many there are. Therefore, they should always add to each other (like partial pressures in a mixture of gases.)

5.62 24.6 atm

5.63 4.40 atm

5.64 A) 0.9053 B) 0.6999

5.65 A) (a) Solution I: $\gamma_{H_2O} = \dfrac{9.239}{\chi_{H_2O} P^*_{H_2O}} = 0.973$ Solution II: $\gamma_{H_2O} = \dfrac{9.239}{\chi_{H_2O} P^*_{H_2O}} = 0.958$

(b) No, it is decreasing with increasing solute concentration as would be expected.

B) Because the value of γ is less than 1.0, the ln(γ) is negative and we would expect H^E for the mixing to be exothermic. It also indicates the raffinose ⋯ water interactions are stronger than in water, indicating some clustering of water and the saccharide is likely occurring.

C) (a) $m = 0.4025$ $\Delta T_{fp,ideal} = -0.749°C$

(b) Given that it likely that the activity coefficient for water will be less than 1.0, the vapor pressure will not be lowered as expected for an ideal solution, therefore the depression observed will be less than $-0.749°C$.

5.66 A) $\gamma_{H_2O} = \dfrac{P_{soln}}{\chi_{H_2O} P^*_{H_2O}} = 0.889$ B) $\Delta T_{fp,obs} = \gamma_A \Delta T_{fp,A} = 0.889(-19.3°C) = \underline{-17.2°C}$

C) Quite a significant impact. The freezing point obtained with the correction for the activity of water is 2°C higher than the value calculated with the first equation, and 6.6°C higher than calculated from the simplified equation

5.67   A)   0.980   B) –2.11°C   C) 25.1 atm
    D)   Not a significant difference (only about 2%) in this case since $\Pi_{\gamma=1.0} = 25.6 \, \text{atm}$.

5.68   A)   0.906   B) 3.41:1.0 so there are about 7 water molecules for every 2 lactose molecules
    C)   The actual mole fraction of lactose is 0.2275, but, if calculated from $\Delta P_A = \chi_B P_A^*$, the value is 0.300, so this solution does illustrate that effect.

5.69   A)   1.06
    B)   The value of the activity coefficient of water is greater than 1.0 instead of less than 1.0, as was found for the other solutions.
    C)   Yes, even though the solutes are non-volatile, it is the strength of the A···B interactions and the possible clustering and grouping of molecules, as we saw earlier, that can affect vapor pressure and produce both values of activity coefficient that are higher than one or less than one. The colligative properties should also reflect what is going on between the molecules in solution.

5.70   A)

Solution	ΔT_{fp} (°C)	$\Delta T_{fp,\gamma=1.0}$	γ_{H_2O}	B) No. "free" water molecules vs. pure water
20.0% ethylene glycol	–7.93	–7.49	1.06	Increased
20.0% 1-propanol	–7.76	–7.74	1.00	Stayed the same
20.0% urea	–7.00	–7.74	0.904	Decreased

    B)   All interactions between water and the solutes are H-bonding, but urea has a flat planar structure with many H-bonding sites, compared to ethylene glycol or propanol, creating clustering within the solution. Ethylene glycol appears to increase the number of free water molecules (as did lactose, another "polyol") so it may be disrupting the normal H-bonding arrangements in water.
    C) $\Pi_{obs} = \gamma_A (\Pi_{\gamma=1.0}) = 0.904 (85.72 \, \text{atm}) = \underline{77.66 \, \text{atm}}$

    You would overestimate the pressure by 10.4% so this is a significant difference.

5.71   7.05 atm

5.72   A) $2.32 \times 10^{-3} \, \text{M} \, CCl_2F_2$ B) K_H, $CCl_2F_2 = 2.32 \times 10^{-3} \, \text{M/atm}$

5.73   A) $C_{heptachlor} = 1.42 \times 10^{-7} \, \text{M}$ $P_{gas} = 1.15 \times 10^{-7} \, \text{atm}$
    B)   No. molecules/cm^3 over water = $\underline{2.76 \times 10^{12}}$
    C)   If K_H is increased then the pressure of the gaseous molecules over the solution would be greater if the concentration of the dissolved gas were the same.
    D)   No. molecules/cm^3 over seawater = $\underline{7.06 \times 10^{12}}$

5.74   A)   2.70 M   B) The molarity calculated from the literature value is 2.80 M, so the values agree.

5.75   A) $\chi_B = 0.473$
    B)   The saturated solution has $\chi_B = 0.132$ so $C_6H_4Cl_2$ is much less soluble than in an ideal solution.
    C)   p-Dichlorobenzene is a non-polar molecule (since Cl's are opposite each other on the ring) with strong LDF forces, largely due to the aromatic ring. Hexane is also non-polar, but its LDF forces would be much weaker, so that A···B forces are weaker than B···B and they don't mix as well and do not form an ideal solution.

5.76 $T = 34°C$

PART 6: FREE ENERGY (ΔG), EQUILIBRIUM, AND ELECTROCHEMISTRY

6.1   A) $K_p = 6.46 \times 10^{-4}$ B) $\Delta G_f^\circ (NH_4)_2 CO_3 (s) = -408.8$ kJ/mol

6.2   A)   (a) +22.6 kJ   (b) $K_p = P O_2$ (c) $K < 1.0$ since $\Delta G^\circ (+)$
     B) $P O_2 = 1.09 \times 10^{-4}$ atm
     C) $T = 172$ K ($-101°C$)   ▪ Since $\Delta H_r^\circ (-)$, must cool reaction to increase $K = P O_2$.

6.3   A) $NO_2(g) \to NO(g) + \frac{1}{2} O_2(g)$ and $K_p = \dfrac{P_{NO(g)} \left(P_{O_2(g)} \right)^{1/2}}{P_{NO_2(g)}}$

     B) $K_p = 0.378$ @ 700 K, $\Delta G^\circ = +5.68$ kJ $K_p = 1.33$ @ 800 K, $\Delta G^\circ = -1.90$ kJ
     C)   Since K increases when T increases, $\Delta H(+)$ and because Δn_{gas} (+) for reaction, expect $\Delta S(+)$.
     D) $\Delta H^\circ = +57.1$ kJ, $\Delta S^\circ = +73.2$ J/K makes T changeover = 780 K

     ▪ This correlates to experimental data, since $\Delta G^\circ (+)$ at 700 K, but $\Delta G^\circ (-)$ at 800 K.

6.4   A)   (a) $K_p = 49.6$
         (b)   The value of K_p would be the same since the conversion factors needed will cancel out in this case since Δn for gases = 0. But if they don't cancel, then it could make a difference in the quoted value for K_p.
     B) $\Delta G_r^\circ = -23.7$ kJ
     C)   ▪ Need to use $\Delta H(T_2) = \Delta H(T_1) + \Delta C_p(\Delta T)$ and $\Delta S(T_2) = \Delta S(T_1) + \Delta C_p \ln(T_2/T_1)$
         $\Delta G_{731} = \Delta H_{731} - T\Delta S_{731} = -23.8$ kJ   So the values agree.

6.5   A) $P_{PCl_5} = 0.136$ atm

     B)   (a) $\% PCl_5 = \dfrac{0.136}{0.894} \times 100 = 15.2\%$

         (b)   Expect that when $K > 1.0$ will have more products than reactants and that is the case here, with the PCl_5 only making up 15.2% of the final mixture.
     C)   Weight PCl_5 left = 1.32 g

6.6   A)   Not at equilibrium, $Q = 0.25 > K = 0.106$ so iso-borneal must be converted to the borneal isomer.
     B)   Wt. borneal at equilibrium = 18.08 g and wt. isoborneal = 1.82 g
     C) $\Delta G = \Delta G^\circ + RT \ln Q = -RT \ln K + RT \ln Q = -3.59$ kJ

6.7   A) $K_C = 542$
     B)   78.6% conversion of limiting reactant, so more product than reactant left at equilibrium as expected when $K > 1.0$. The % conversion of amylene appears low, but it was in excess and does not limit product formation.

6.8   A) $K_{P,693\,K} = 0.0180$, $K_{P,723\,K} = 0.179$
     B) $K_{C,693\,K} = 9.79 \times 10^{-8}$, $K_{P,723\,K} = 8.59 \times 10^{-7}$
     C)   There would be some difference, since the log terms would be different. It's mostly due to rounding-off errors – since it is only about a 5% difference.
     D)   Should quote average: $\Delta H_r = 310$ kJ (± 9.0 kJ)

6.9   A) $\Delta G_r^\circ = +27.0$ kJ
     B) $K_c = 1.85 \times 10^{-5}$, literature value 1.8×10^{-5} so ion values appear accurate.
     C)   ▪ Will need the value of ΔH and ΔS for the reaction to determine their signs.
         $\Delta H_r^\circ = +3.59$ kJ, $\Delta S_r^\circ = -78.6$ J/K mol so that ΔG° is positive at all temperatures.

6.10   A)   *See the Full Solutions for plot and data*
      B) $\Delta H_r^\circ = +58.9$ kJ/mol
      C)   Use equation from plot: $\Delta G_r^\circ = +12.6$ kJ/mol $\Delta S_r^\circ = 158$ J/K mol

6.11   A)   (a) $K_p = 2.25$ $\Delta G^\circ = -4.54$ kJ   (b) $\Delta H_r^\circ = 86.8$ kJ $\Delta S_r^\circ = 144.6$ J/K
      B) $K_c = 0.0407$ C) ▪ Will need to use successive approximations. $[HI]_{equil} \approx 0.069$ M

6.12 A) Q versus K question: $Q = 1.32$ whereas $K = 0.00552$ (from $\Delta G°$) so the reverse reaction rate will dominate and products will be converted to reactants.

6.13 A) Coupled reaction: $PEP + ADP \Leftrightarrow Py + ATP$, $\Delta G° = -24.3$ kJ, $K = 1.82 \times 10^4$.
B) $[ATP] = 0.00993$, % conversion $= (0.00993)/(0.01) \times 100 = 99.3\%$
C) Without coupling reaction (1) $K = 2.49 \times 10^9$ reaction (2), $K = 7.31 \times 10^{-6}$
D) Without coupling, the percent conversion of ADP to ATP would have been much lower, approximately 2.7%, because of the low value of K. The 99% conversion of ADP would have been impossible without coupling.

6.14 A) $K = \dfrac{[\text{alanine}][\text{oxolacetate}]}{[\text{pyruvate}][\text{asparate}]} = 0.221$ Given $\Delta G_r° = 8308$ J

B) ▪ Must calculate K and then ΔG: $Q = 1.0 \times 10^{-5}$ $\Delta G_r = \Delta G_r° + RT \ln Q = -25.2$ kJ
▪ So reaction will be spontaneous under the cell conditions.

6.15 A) Both FAD and NADH are the reactants, so the second reaction must be reversed before adding.

$$\cancel{2e^-} + FAD + \cancel{2}H^+ \rightarrow FADH_2$$

$$+ (NADH \rightarrow NAD^+ + \cancel{H^+} + \cancel{2e^-}) \qquad \Delta G°_{\text{overall}} = (42.3 + (-65.5)) = \underline{\underline{-23.2 \text{ kJ}}}$$

$$\overline{H^+ + FAD + NADH \rightarrow FADH_2 + NAD^+}$$

B) $\ln K_c = 9.364$ $K_c = e^{9.364} = \underline{1.17 \times 10^4}$
C) The K_c term involves [H⁺], so the value of all equilibrium concentrations will depend on the pH. This includes the thermodynamic values as well. Therefore, to create a consistent reference value, the pH must be set for biochemical reactions, like those involved in this example, and that has been determined to be pH = 7.0 as part of the "biochemical standard state" conditions.

6.16 A) $\Delta G°_{\text{STEP I}} = 8.3$ kJ   B) [citrate] $= 0.342$ M
C) $\Delta G°_{\text{STEP I}} = 5.0$ kJ   D) $K_{\text{STEP II}} = 0.133$

6.17 A) Equation from plot: $y = -56.838x - 1.6645$, then:
$\Delta H° = -R(\text{slope}) = 473$ J/mol and $\Delta S° = R(y\text{-intercept}) = -13.8$ J/K mol
B) $\Delta G° = 4.60$ kJ/mol   C) [2-Pgly] $= 0.20$ M, [3-Pgly] $= 0.130$ M

6.18

	Dilute Solution or pure water:	Higher ionic strength solutions:
A	$K_a, HA = \dfrac{[H_3O^+][A^-]}{[HA]}$	$K_a, HA = \dfrac{a_{H_3O^+} a_{A^-}}{[HA]} = \dfrac{\gamma_{H_3O^+}[H_3O^+] \times \gamma_{A^-}[A^-]}{[HA]} = \left(\gamma_{H_3O^+} \gamma_{A^-}\right) \times \dfrac{[H_3O^+][A^-]}{[HA]}$
B	$K_b, B = \dfrac{[BH^+][OH^-]}{[B]}$	$K_b, B = \dfrac{a_{BH^+} a_{OH^-}}{[B]} = \dfrac{\gamma_{BH^+}[BH^+] \times \gamma_{OH^-}[OH^-]}{[B]} = \left(\gamma_{BH^+} \gamma_{OH^-}\right) \times \dfrac{[BH^+][OH^-]}{[B]}$
C	$K_f, [Fe(H_2O)_6]^{+2} = \dfrac{[Fe(H_2O)_6]^{+2}}{[Fe^{+2}]}$	$K_f, [Fe(H_2O)_6]^{+2} = \dfrac{a_{[Fe(H_2O)_6]^{+2}}}{a_{[Fe^{+2}]}} = \dfrac{\gamma_{[Fe(H_2O)_6]^{+2}}}{\gamma_{[Fe^{+2}]}} \times \dfrac{[Fe(H_2O)_6]^{+2}}{[Fe^{+2}]}$
D	$K_f, [CoCl_4]^{-2} = \dfrac{[CoCl_4]^{-2}}{[Co^{+2}][Cl^-]^4}$	$K_f, [CoCl_4]^{-2} = \dfrac{a_{[CoCl_4]^{-2}}}{a_{Co^{+2}}(a_{Cl^-})^4} = \dfrac{\gamma_{[CoCl_4]^{-2}}}{\gamma_{Co^{+2}}(\gamma_{Cl^-})^4} \times \dfrac{[CoCl_4]^{-2}}{[Co^{+2}][Cl^-]^4}$

6.19 (a) $K_{sp} = a_{Ca^{+2}} a_{CO_3^{-2}} = \gamma_{Ca^{+2}} \gamma_{CO_3^{-2}} [Ca^{+2}][CO_3^{-2}]$

The values for the activity coefficients will be less than 1.0, so the product will be a decimal value. In order to equal the same K_{sp} value (set by the thermodynamics), the solubility or dissolved concentrations of ions must increase.

(b) $K_b = \dfrac{a_{BH^+} a_{OH^-}}{a_B} = \gamma_{BH^+} \gamma_{OH^-} \left(\dfrac{[BH^+][OH^-]}{[B]} \right)$

The values for the activity coefficients for the two ions, BH+ and OH-, will be less than 1.0 but that for the neutral weak base, B, will be 1.0. Consequently, in order to keep K the same, the ratio of ions to neutral weak base at equilibrium must increase.

B) $K_f = \dfrac{a_{[FeSCN^{+2}]}}{a_{Fe^{+3}} a_{SCN^-}} = \dfrac{\gamma_{[FeSCN^{+2}]}}{\gamma_{Fe^{+3}} \gamma_{SCN^-}} \left(\dfrac{[FeSCN^{+2}]}{[Fe^{+3}][SCN^-]} \right) = \dfrac{\gamma_{[FeSCN^{+2}]}}{\gamma_{Fe^{+3}} \gamma_{SCN^-}} (Q_{complex\,to\,ions})$

As the ionic strength increases, the ratio of activity coefficients will get smaller since there are two ion terms in the denominator, but only one in the numerator. So, assuming the activities are close in value, the effective ratio is $1/\gamma$ and gets larger as I increases, since γ will get smaller. Conse-quently, Q, the ratio of complex to ions, should get smaller as I increases, so that the product K_f stays the same.

6.20 A) 1.45×10^{-5} in pure water   B) 4.0×10^{-5} in 0.10 M NaCl

6.21 A)   *See the Full Solutions for proof.*
       B)   (a) pH = 2.87 for dissociation in water   (b) pH = 2.20 for dissociation in 1.0 M KCl

6.22 A)   *See the Full Solutions for proof.*
       B)   (a) pH = 7.34 ignoring activities   (b) pH = 6.58 including activities

6.23 A) $\Delta E° = 1.17$ V if ClO_4^- reduced.
       B)   Overall Reaction: $ClO_4^- + Cd(s) + H_2O(l) \rightarrow Cd(OH)_2(s) + ClO_3^-$ and $K = [ClO_3^-]/[ClO_4^-]$
       C)   Cell notation: $Cd| Cd(OH)_2(s)| OH^-||OH^-, ClO_3^-, ClO_4^-|Pt$
       D)   Since $n = 2$ mol e's in balanced reaction, then $\Delta G° = -226$ kJ, ln $K = 91.14$ and $K = e^{91.14} = 3.82 \times 10^{39}$
       E)   Since the K value is so high, it is very likely that the ratio of ClO_3^- to ClO_4^- ions could not be determined by any direct measurements since ClO_4^- would always be extremely small at equilibrium.

6.24   ■ Work $= -nF\Delta E° = \Delta G°$ for overall reaction, then calculate per mole of metal oxidized.
       Answers: Reactions a, c

6.25 A)   Need to have the overall reaction be: $Hg_2SO_4(s) \Leftrightarrow Hg_2^{+2} + SO_4^{-2}$
       Cathode: $Hg_2SO_4(s) + 2e^- \rightarrow 2\,Hg(l) + SO_4^{-2}$ $E° = 0.6125$ V
       Anode: $Hg_2^{+2} + 2e- \rightarrow 2\,Hg(l)$ $E° = 0.7973$ V $\Delta E° = -0.1848$ V

       B)   ln $K_{sp} = -14.395$ so K_{sp} $Hg_2SO_4(s) = e^{-14.395} = 5.60 \times 10^{-7}$

       C)   The literature value of 6.5×10^{-7} is quite close the calculated value. A difference of a few hundredths of a volt in the $\Delta E°$ could account for the difference.

6.26 A)   Reverse the first half reaction (so it becomes anode) and add it to the second half reaction (cathode) to get: $Hg^{+2} + 4\,CN^- \Leftrightarrow Hg(CN)_4^-$
       B) $\Delta E° = 1.23$ V so $\Delta G° = -23.7$ kJ then $K_{formation} = e^{95.8} = 4.09 \times 10^{41}$
       C)   (a) $Hg(l)|Hg(CN)_4^{-2}, CN^-||Hg^{+2}|Hg(l)$
              (b) Need an electrical connection to the Hg(l) even though it can act as the electrode.

6.27 A) $\Delta E° = 0.18$ V   B) $K = 1.22 \times 10^6$ C) (a) $\Delta E = 0.15$ V, (b) Voltage is the same since Q is the same
       D)   If the same amount of HCl is added to both cells, the cell voltage should stay the same.

6.28 A) ▪ Use Nernst equation but apply to half reaction.
If assume standard conditions for other ions, $[Mn^{+2}] = [MnO_4^{-2}] = 1.0$ M then at:
(a) pH = 6.0 $E_{MnO_4^-/Mn^{+2}} = \underline{0.942\ V}$ (b) pH = 2.0 $E_{MnO_4^-/Mn^{+2}} = \underline{1.32\ V}$

B) The lower pH produces a more positive value of E, so when coupled in a redox reaction, you should favor getting more product.

6.29 Combustion/Fuel cell reaction: $CH_4(g) + 2\,O_2(g) \rightarrow CO_2(g) + 2\,H_2O(l)$
▪ $CH_4 \rightarrow CO_2$ *loss of 8 electrons* ($-4 \rightarrow +4$ *change in oxidation number*)
▪ *4 O atoms gaining 2 electrons each* ($0 \rightarrow -2$ *change in oxidation number*)

A) ▪ Can get $\Delta E°$ from $\Delta G°$ for reaction.

$\Delta G° = -818$ kJ. Then given $n = 8$ mole electrons: $\Delta E° = \underline{1.059\ V}$
Yes, this fuel cell produces just a little over 1.0 volt at 25°C.

B) Half-reaction for CH_4: $CH_4(g) + 2\,H_2O(l) \rightarrow CO_2(g) + 8\,H^+ + 8\,e^-$

$$\Delta E° = E°_{O_2/H_2O} - E°_{CH_4/CO_2} \Rightarrow E°_{CH_4/CO_2} = \underline{0.17\ V}$$

C) Need moles CH_4 reacted:

$$n = 2.45\,\text{mol} \left[\frac{-818\text{kJ}}{1\,\text{mol}\,CH_4} \right] = -2008\text{kJ} = \Delta G = w_{max}$$

6.30 Cathode: $Fe^{+3} + e^- \rightarrow Fe^{+2}$ $E° = 0.77$ V   Anode: $Cu^{+2} + 2e^- \rightarrow Cu(s)$ $E° = 0.34$ V
Overall reaction: $2\,Fe^{+3} + Cu(s) \rightarrow 2Fe^{+2} + Cu^{+2}$ $\Delta E° = 0.43$ V

$$\Delta E = \Delta E° - \frac{0.0257}{n} \ln Q \Rightarrow \ln Q = -3.268$$

$$Q = \frac{[Cu^{+2}][Fe^{+2}]^2}{[Fe^{+3}]^2} = \frac{[Cu^{+2}][0.060]^2}{[0.14]^2} \Rightarrow [Cu^{+2}] = \underline{0.207\ M}$$

6.31 A) *See the Full Solutions for derivation.*
B) (a) Highest = I   (b) Lowest = III   (c) Same II, IV

6.32 $\left[\dfrac{[K^+]_{inside}}{[K^+]_{outside}} \right] = 10^{-1.133} = \underline{0.0736}$

6.33 $\Delta E = 0.0592\left(pH_{anode} - pH_{cathode}\right)$

$E_{cathode} - E_{anode} = 84$ mV $= 59.2$ mV $\left(pH_{anode} - 3.00\right) \Rightarrow \underline{4.42} = pH_{unknown}$

6.34 ▪ Given ΔE, not $\Delta E°$, so can't calculate K from $\Delta E°$. Must find $[Cu^+]$ on right side since that equals "s", the molar solubility of CuCl(s) and then apply $K_{sp} = s^2$.

$$\Delta E = 0.175V = 0.0592 \log \left[\frac{[Cu^+]_{Left}}{[Cu^+]_{Right}} \right] \Rightarrow \left[\frac{1.00\ M}{[Cu^+]_{Right}} \right] = 10^{2.956}$$

$$\Rightarrow [Cu^+]_{Right} = s = 1.106 \times 10^{-3}\ M$$

$$K_{sp} = \left(1.106 \times 10^{-3}\right)^2 = \underline{1.22 \times 10^{-6}}$$

6.35 A) $\Delta E°$ must be positive to produce a spontaneous reaction
$\Delta E° = E°_{cathode(red)} - E°_{anode(oxid)} = -0.320 - (-0.42)$ V $= \underline{0.10\ V}$ so NAD^+ reduced.

B) $\Delta G° = -19.3$ kJ (pH = 7.0) $K = 2416$ (pH = 7.0)

6.36   Cathode: $NAD^+ + 2e^- + H^+ \rightarrow NADH$ Anode: $O_2(g) + 2H^+ + 2e^- \rightarrow H_2O_2$

B) $Pt|O_2(g)$ (P = 1.0 atm)$|H^+, H_2O_2||NADH, H^+, NAD^+|Pt$

C)   (a) $\Delta E'^\circ = 0.615$ V   (b) $\Delta G'^\circ = -314$ kJ   (c) $\ln K' = 126.7$ $K'_{298} = e^{126.7} = 1.0 \times 10^{55}$

6.37   A) $\Delta E'^\circ = -0.219 - (-0.320)$ $V = 0.101$ V so $\Delta G'^\circ = -19.5$ kJ

B)   Since the hydrolysis can take 30.5–35.0 kJ depending on pH, this is not enough energy, providing only about 2/3 of what is needed.

6.38   A)   Balanced reaction: 2 $cytochrome$ $c(Fe^{+3}) + lactate$ $\Leftrightarrow$ 2 $cytochrome$ $c(Fe^{+2}) + pyruvate$

Then: $\Delta E'^\circ = 0.254 - (-0.185)$ V $= 0.439$ V

B)   *See the Full Solutions for derivation.* $K' = 6.47 \times 10^{14}$

C)   *[pyruvate]/[lactate]* $= 6.47 \times 10^8$

D) $Q < K$ so [pyruvate] increases. *See the Full Solutions for proof.*

6.39   A)   See table of results on the right

B) $\dfrac{K_{pH=7.0}}{K_{pH=5.0}} = 1.00 \times 10^6$

Half reaction (298 K)	pH=5.0	pH=6.0	pH=7.0
E°', NAD+/NADH (V)	−0.2569	−0.2865	−0.316
E°', Xylitol/xylulose (V)	0.1076	0.1667	0.2259
$\Delta E^\circ = E^\circ_{cathode} - E^\circ_{anode}$	**0.3645**	**0.4532**	**0.5419**
ΔG°' (kJ)	**70.35**	**87.47**	**104.59**

C)   (a)   24.4%   (b) 19.5%
(c)   48.7%

6.40   A)   Applying the standard thermodynamics calculation:

$$\Delta G'^\circ = \left[2\,mol\left(\Delta G'^\circ_{f,cytochrome\text{-}c(Fe^{+2})}, \frac{kJ}{mol} \right) + 1\,mol\left(\Delta G'^\circ_{f,pyruvate}, \frac{kJ}{mol} \right) \right]$$
$$- \left[2\,mol\left(\Delta G'^\circ_{f,cytochrome\text{-}c(Fe^{+3})}, \frac{kJ}{mol} \right) + 1\,mol\left(\Delta G'^\circ_{f,lactate}, \frac{kJ}{mol} \right) \right]$$

At $I = 0$:   A) $\Delta G'^\circ = -84.5$ kJ   B) (a) Use $\ln K' = -\dfrac{\Delta G'^\circ}{RT}$ then $K = 6.6 \times 10^{14}$

At $I = 0.10$:   A) $\Delta G'^\circ = -79.6$ kJ   B) (a) $K = 9.0 \times 10^{13}$

At $I = 0.25$:   A) $\Delta G'^\circ = -78.0$ kJ   B) (a) $K = 4.8 \times 10^{13}$

B)   They agree very well: 6.6×10^{14} versus 6.74×10^{14} from $\Delta E'^\circ$.

C) K' is 7.33 times larger at $I = 0$ than at $I = 0.10$, while K' is 13.8 times larger at $I = 0$ than at $I = 0.25$, so K is decreasing with ionic strength in this reaction.

D)   Cellular environments have higher values of ionic strength so that knowing how K' depends on ionic strength gives a better picture of the reaction in its real environment.

6.41   A)   Balanced reaction: $NAD^+ + CH_3CH_2OH \Leftrightarrow NADH + CH_3CHO + H^+$

B)   Applying the standard thermodynamics calculation:

$$\Delta G'^\circ = \left[1\,mol\left(\Delta G'^\circ_{f,NADH}, \frac{kJ}{mol} \right) + 1\,mol\left(\Delta G'^\circ_{f,CH_3CHO}, \frac{kJ}{mol} \right) \right]$$
$$- \left[1\,mol\left(\Delta G'^\circ_{f,NAD^+}, \frac{kJ}{mol} \right) + 1\,mol\left(\Delta G'^\circ_{f,ethanol}, \frac{kJ}{mol} \right) \right]$$

At $I = 0$: $\Delta G'^\circ = +25.3$ kJ   leads to: $K' = 3.72 \times 10^{-5}$

At $I = 0.25$: $\Delta G'^\circ = +22.1$ kJ   leads to: $K' = 1.33 \times 10^{-4}$

C)   In this reaction, K is increasing with increasing ionic strength, opposite to the situation in Problem 6.40.

D) (a) $K' = K_c \times 10^{7x}$ In this reaction $x = +1.0$ since H^+ is a product, so that:

$$K_c = \frac{K'}{10^7} = \underline{3.72 \times 10^{-12}}$$

(b) If $[H^+] = 1.0$ M for K_c and 1.0×10^{-7} for K' then a shift towards product should occur since the numerator in Q has decreased dramatically, so it is consistent.

6.42 A) Balanced reaction: $O_2(g) + 2 \; cysteine \Leftrightarrow 2 \; H_2O(l) + 4 \; cystine$, $\Delta E'^\circ = 1.16$ V, $\Delta G'^\circ = -446$ kJ

B) The concentration should decrease as the $O_2(g)$ converts cysteine spontaneously to cystine in the very product-favored reaction.

6.43 A) (a) $2 \; AgCl(s) + Zn(s) \rightarrow 2 \; Ag(s) + 2 \; Cl^- + Zn^{+2}$ (b) $\Delta E^\circ = (0.22 + 0.76) = 0.98$ V

(c) (1) AgCl(s) needed for cathode reaction is coating the electrode at the bottom

(2) Ag(s) needed for cathode reaction and functions as one of the metal electrodes

(3) Zn(s) needed for the anode reaction and functions as the second metal electrode

(4) Solution of $ZnCl_2$, containing Zn^{+2} ions needed for anode reaction

(5) The function of this part is for the electrical connections needed to measure voltage.

(d) Since the reactants are both solids, they cannot come into contact and exchange electrons except through the external wires so we don't need to separate them.

B) ◼ Must define Q in terms of $[Zn^{+2}]$ and then solve for Q from Nernst Equation:

$$[Zn^{+2}] = x, [Cl^-] = 2x \text{ leads to: } Q = [Zn^{+2}][Cl^-]^2 = (x)(2x)^2 = 4x^3$$

$$\Delta E - \Delta E^\circ = -\frac{0.0257}{n} \ln Q \Rightarrow \ln Q = -2.724$$

$$Q = 0.06563 = 4x^3 \Rightarrow x = [Zn^{+2}] = \sqrt[3]{0.01641} = \underline{0.254 \text{ M}}$$

C) $\Delta G^\circ = -nF\Delta E^\circ = \underline{-189 \text{ kJ}}$ $\Delta S^\circ = nF\left[\dfrac{d\Delta E^\circ}{dT}\right]_P = \underline{-77.6 \text{ J/K}}$ $\Delta H^\circ = \Delta G^\circ + (T\Delta S^\circ) = \underline{-218 \text{ kJ}}$

6.44 Overall reaction: $Hg_2Cl_2(s) + H_2(g) \Leftrightarrow Hg(l) + 2 \; HCl$ $n = 2$ mol electrons

◼ Need to interpolate value of ΔE° at 298 K from data to calculate ΔG° at 298 K.

◼ Estimate derivative as: $(\Delta E^\circ_{303} - \Delta E^\circ_{303})/(303 - 293)$ K.

$$\Delta E^\circ_{298} = \Delta E^\circ_{293} - \left(\frac{5K}{10K}\right)(\Delta E^\circ_{303} - \Delta E^\circ_{293}) = 2.6975 \text{ V}$$ $$\Delta G^\circ = -nF\Delta E^\circ_{298K} = \underline{-520.6 \text{ kJ}}$$

$$\Delta S^\circ \approx nF\left[\frac{\Delta(\Delta E^\circ)}{\Delta T}\right]_P = \underline{-57.9 \text{ J/K}}$$ $$\Delta H^\circ = \underline{-537.8 \text{ kJ}}$$

6.45 A) Cathode: $Hg_2Cl_2(s) + 2e^- \rightarrow 2 \; Hg(l) + 2 \; Cl^-$ $E^\circ = 0.268$ V

Anode: $AgCl(s) + e^- \rightarrow Ag(s) + Cl^-$ $E^\circ = 0.222$ V

Overall: $Hg_2Cl_2(s) + 2 \; Ag(s) \Leftrightarrow 2 \; Hg(l) + 2 \; AgCl(s)$

B) $\Delta E^\circ = 0.046$ V $\Delta G^\circ = -8878$ J $= \underline{-8.88 \text{ kJ}}$

C) ◼ Must plot data, as ΔE(volts) versus T(K)   From equation: slope $= 3.544 \times 10^{-4}$ V/K

$$\Delta S^\circ = nF\left[\frac{d\Delta E^\circ}{dT}\right]_P = 2.0 \text{ mol}\left(9.65 \times 10^4 \frac{\text{coul}}{\text{mol}}\right)\left(3.54 \times 10^{-4} \frac{\text{V}}{\text{K}}\right) = \underline{68.4 \text{ J/K}}$$

$$\Delta H^\circ = \Delta G^\circ + (T\Delta S^\circ) = \underline{11.5 \text{ kJ}}$$

D) Since in this cell $\Delta H°(+)$ and $\Delta S°(+)$, then $\Delta G°$ will change sign with T, as will $\Delta E°$. It should be spontaneous at higher T's, non-spontaneous at lower T's. Since the $\Delta E°$ is positive we are <u>above</u> the changeover temperature at 298 K.

6.46 A) Calculated values:
B) *See the Full Solutions for plot.* $\Delta S° = -60.0$ J/K
C) *See the Full Solutions for plot.* $\Delta H° = -39.25$ kJ
D) *See the Full Solutions for calculated results.* There is excellent agreement between the values of ΔH calculated from each $\Delta E°$ ranging from $-39.26 \rightarrow -39.32$ kJ.

T (K)	ΔE°	ΔG°	ΔG° kJ	ln K	K
280	0.23302	−22486	−22.49	9.6595	15669.2
284	0.23085	−22277	−22.28	9.4347	12515.4
288	0.22857	−220557	−22.06	9.2118	10014.5
292	0.22619	−21827	−21.83	8.9910	8030.5
296	0.22371	−21588	−21.59	8.7722	6452.7
300	0.22112	−21338	−21.34	8.5551	5193.1
304	0.21843	−21078	−21.08	8.3398	4187.3
308	0.21564	−20809	−20.81	8.1264	3382.5

6.47 A) Balanced reaction: $AgBr(s) + \frac{1}{2} H_2(g) \rightarrow Ag(s) + HBr(aq)$
and $K_{dissociation} = \dfrac{[H^+][Br^-]}{[HBr]}$

B) (a) Calculated results for $\Delta G°$: *See the Full Solutions*
(b) As T increases, $\Delta G°$ for the dissociation decreases, indicating the dissociation is less spontaneous at the higher T's. This indicates the sign of ΔH is negative for the dissociation.
(c) When the % ethanol increases, the $\Delta G°$ decreases indicating the change in solvent is impeding the dissociation. So the dielectric strength of the solvents is affecting the dissociation of the acid. In addition, the activity effects in the binary mixture discussed earlier in Part 5 (page 45, 47), may also be playing a role in impeding the dissociation.

C) (a) Calculated results for K_c for the dissociation: *See the Full Solutions*
(b) In pure water, the K values, although greater than 1.0, decrease dramatically as T increases. The same trend is observed in the other two solvents. The ratio of the greatest value of K_{diss} to the least is about 3 in each solvent. This would indicate that the ΔH values for the dissociation are similar for each solvent.

D) Plot of data for $\Delta S°$: *See the Full Solutions*
(a) The results are not linear equations, but would be polynomials in T. So $\Delta S°$ is not staying constant over the T range studied.
(b) Estimating the $\Delta S°$ value within each 10 K region using: *See the Full Solutions*
The estimated $\Delta S°$ values change over the T regions, becoming more negative at the higher T's, with some exceptions for parts of the 20% and 50% ethanol data.
From the graph, one can observe that the 50% ethanol produces larger shifts in $\Delta S°$ as the T is increased and that is also shown in the estimated values. Considering the interactions of ethanol and water are quite strong in this weight % region (50% ethanol by mass = mol fraction of 0.28 for ethanol), it is not surprising the entropy for the dissociation of the ions is being affected.

6.48 ▪ Need to calculate ionic strength: $I = 0.0150$ m and $\sqrt{I} = 0.1225$
A) $\log \gamma_\pm = -0.111$ then $\gamma_\pm = 10^{-0.111} = \underline{0.774}$, $m_\pm = 7.94 \times 10^{-3}$ m and $a_\pm = \gamma_\pm m_\pm = \underline{6.14 \times 10^{-3}}$
B) $I = 0.0350$ m and $\sqrt{I} = 0.187$ then $\log \gamma_\pm = -0.1604$ and $\gamma_\pm = 10^{-0.1604} = \underline{0.691}$
For activity, $m\pm$ stays the same, so that $a_\pm = \gamma_\pm m_\pm = \underline{5.49 \times 10^{-3}}$ m
C) Using the Davies equation: $\log \gamma_\pm = -0.171$ $\gamma_\pm = 10^{-0.171} = \underline{0.674}$ and $a_\pm = \underline{5.36 \times 10^{-3}}$ m
No, although the coefficients show some difference, the activities are very much the same.

6.49 A) $I = \dfrac{1}{2} \sum_{all\ ions} c_i z_i^2 = \dfrac{1}{2}\left[0.40(+1)^2 + 0.20(-2)^2\right] = 0.600$ m so $\sqrt{I} = 0.775$

$$\log \gamma_\pm = -0.444 \quad \text{and} \quad \gamma_\pm = \underline{0.360} \quad m_\pm = (m_+^{v+} m_-^{v-})^{1/v} = 0.252 \text{ m} \quad a_\pm = \gamma_\pm m_\pm = 0.360$$
$$(0.252) = \underline{0.0907 \text{ m}}$$

B) Davies: $\log \gamma_\pm = -0.624$ and $\gamma_\pm = 10^{-0.624} = \underline{0.238}$ $a_\pm = \gamma_\pm m_\pm = 0.238(0.252) = \underline{0.0599 \text{ m}}$

6.50 A) Results of calculation:

B) For $BaBr_2$, both the Debye–Hückel and Davies equations predict well up to 0.010 M. After that, the equations do not predict well.

For CsF, both equations predict well up to 0.10 M and they are both close to the actual value. Above 0.50 M the equations overestimate the coefficient.

m (mol/kg)	0.001	0.005	0.010	0.100	0.500	1.000
			$\gamma\pm$ BaBr$_2$			
Actual	0.881	0.785	0.727	0.517	0.435	0.470
$\gamma\pm$ Debye-Huckel	0.885	0.774	0.707	0.436	0.275	0.226
$\gamma\pm$ Davies	0.887	0.783	0.723	0.539	0.790	1.866
			$\gamma\pm$ CsF			
Actual	0.965	0.929	0.905	0.792	0.721	0.726
$\gamma\pm$ Debye-Huckel	0.965	0.926	0.899	0.755	0.615	0.557
$\gamma\pm$ Davies	0.965	0.929	0.905	0.810	0.875	1.124

6.51 A) $m_\pm \text{ NaBrO}_3 = (m_+^{v+} m_-^{v-})^{1/v} = ((0.50) \times (0.50))^{1/2} = \underline{0.50}$

$a_\pm, \text{NaBrO}_3 = \gamma_\pm m_\pm = \underline{0.303 \text{ m}}$

Activity of the compound, $NaBrO_3$: $a = a_\pm^v = (.3025)^2 = 0.0915 \text{ m}^2$

B) $m_\pm \text{ CuBr}_2 = (m_+^{v+} m_-^{v-})^{1/v} = ((0.20) \times (0.40)^2)^{1/3} = \underline{0.317 \text{ m}}$

$a_\pm, \text{CuBr}_2 = \gamma_\pm m_\pm = \underline{0.166 \text{ m}}$

Activity of the compound: $a = a_\pm^v = (.166)^3 = \underline{4.56 \times 10^{-3} \text{ m}^3}$

6.52 A) m $BaCl_2 = 01.778$ m   B) $I = \underline{5.34 \text{ m}}$ and $\sqrt{I} = 2.31$

C) $K_{sp} = a_\pm^3 = \gamma_\pm^3 m_\pm^3$ $m_\pm \text{ BaCl}_2 = \underline{2.823 \text{ m}}$ $K_{sp} = 176.94 = \gamma_\pm^3 (2.823)^3$ then $\gamma_\pm = \underline{2.00}$

6.53 Cell reaction: $\frac{1}{2}H_2(g) + AgCl(s) \Leftrightarrow HCl(aq) + Ag(s)$ then n = 1, $\Delta E° = 0.2223$ V
$\Delta E = \Delta E° - 0.0257 \ln(a_\pm^2)$ where $a_\pm^2 = a_{H^+} a_{Cl^-}$ and $\Delta E - \Delta E° = 0.0.0188 \text{ V} = -0.0257(2)\ln(\gamma_\pm m_\pm)$

$m_\pm \text{ HCl} = ((1.0 \text{ m}) \times (1.0 \text{ m})^2)^{1/2} = 1.00 \text{ m}$. Then $\ln(\gamma_\pm m_\pm) = -0.365 = \ln(\gamma_\pm)$ and $\gamma_\pm = e^{-0.365} = \underline{0.694}$

6.54 Cell reaction: $Hg_2Cl_2(s) + Zn(s) \Leftrightarrow ZnCl_2(aq) + 2 Hg(l)$ then $n = 2$, $\Delta E° = 1.0304$ V
So: $\Delta E = \Delta E° - 0.0257 \ln(a_\pm^3)$ where $a_\pm^3 = a_{Zn^{+2}}(a_{Cl^-})^2 = \gamma_\pm^3 m_\pm^3$

Given: $m_\pm \text{ ZnCl}_2 = ((0.0050 \text{ m}) \times (0.010 \text{ m})^2)^{1/3} = 7.94 \times 10^{-3} \text{ m}$ then

$$\Delta E - \Delta E° = 1.2272 \text{ V} - 1.0304 \text{ V} = -0.0257 \left(\frac{3}{2}\right) \ln(\gamma_\pm m_\pm)$$

$$= 0.03855 \left(\ln \gamma_\pm + \ln(7.94 \times 10^{-3}) \right)$$

$$\ln \gamma_\pm = \frac{0.1968 - 0.1864}{-0.03855} = -0.2688 \text{ Then } \gamma_\pm = e^{-0.2688} = \underline{0.764}$$

6.55 A) Overall reaction: $2 AgCl(s) + Cd(Hg) \Leftrightarrow 2 Ag(s) + CdCl_2(aq)$
Cell notation: Cd(Hg)|CdCl$_2$ (aq)|AgCl(s)|Ag

B) $\Delta E = \Delta E° - 0.0257 \ln(a_\pm^3) \Rightarrow \Delta E = \Delta E° - \frac{0.0257(3)}{2} \ln(a_\pm)$ where $a_\pm = \gamma_\pm m_\pm$

$\Delta E° = 0.5732$ V, $n = 2$ and $m_+ \text{CdCl}_2 = ((0.010 \text{ m}) \times (0.020 \text{ m})^2)^{1/3} = 1.587 \times 10^{-4} \text{ m}$

$\Delta E = 0.5732 V - 0.03855 \text{ V} \ln(\gamma_\pm m_\pm) = \underline{0.7478 \text{ V}}.$

Index